PI-3

# PROGRAMMED INSTRUCTION HANDBOOK

## *NONDESTRUCTIVE TESTING*

# *Magnetic Particle*

### *VOLUME I – BASIC PRINCIPLES*

Published by *PH D*iversified, Inc.
5040-B Highway 49 South
Harrisburg, NC 28075

Copyright © 1997 by
*PH D*iversified, Inc.
Second Printing April 2000

All Rights Reserved

No part of this book may be reproduced in any form
without written permission from the publisher.

Printed in the United States of America

ISBN 1-886630-02-X

# PREFACE

**Programmed Instruction Handbook** - Magnetic Particle PI-3 (two volumes) is one of a series of training handbooks designed for **self-study**.

This Programmed Instruction Handbook is also very helpful when used prior to, or in conjunction with, the **Classroom Training Handbook** CT-3, Magnetic Particle. The instructor can make assignments in the classroom handbook and, as a supplement, the student can read corresponding information in this self-study handbook.

This **Programmed Instruction Handbook** presents essentially the same entry-level material as in the Classroom Training Handbook. However, this **self-study format** provides self-evaluation quiz questions at the end of each chapter and at the end of the book. A score of 80 percent or better on these self-tests will indicate a basic understanding of the Level I concepts presented. The **Programmed Instruction and the Classroom Training Handbooks** can both be used successfully as refresher material for Level II training provided that sufficient industry specific equipment, specifications and applications are added to the course outline.

Other **Programmed Instruction Handbooks** in the series include:

| | |
|---|---|
| PI-1 | Introduction to Nondestructive Testing |
| PI-2 | Liquid Penetrant Testing |
| PI-4 | Ultrasonic Testing |
| PI-5 | Eddy Current Testing |
| PI-6 | Radiographic Testing |

It is recommended that PI-1, Introduction to Nondestructive Testing, be completed before starting this book in order to have the benefit of basic metallurgical information that will make this book easier to understand.

# ACKNOWLEDGMENTS

**Publishing and Printing**

    Revision Editor: . . . . . . . Dr. George Pherigo, **PH D**iversified, Inc.

    Production Editor . . Ms. Mary Lou Hollifield, **PH D**iversified, Inc.

    Proofreading . . . . . . . . . . Ms. Jean Pherigo, **PH D**iversified, Inc.
    Proofreading . . . . . . . . . . . . . . . . . . . . . . . . . Ms. Dana Smilie

**Technical Content Revision**

    Technical Editor . . . . . . . . . . . . . . . . . . . Mr. Robert W. Smilie

This handbook was originally prepared by the Convair Division of General Dynamics Corporation under contract to NASA and was identified as N68-28778 This book is part of a series of books, commonly known as the General Dynamics Series, that has been the basis of many industrial NDT training programs for over 20 years.

Now, after several decades of widespread use, the entire series has undergone a major revision. The revised material no longer concentrates on applications in the aerospace industry, but instead, covers a wider range of industrial applications and discusses the newest techniques and applications.

Mr. Robert W. Smilie has been the principal technical editor of the revised material in this text. Using his nondestructive testing experiences in several industries, including work at the EPRI NDE Center, he has updated the text to better suit the entry-level technician/engineer.

# TABLE OF CONTENTS

Chapter 1 - Magnetism ............................... 1-1

   Theory ............................................. 1-1
      Magnetic Attraction and Repulsion ............. 1-3
      Domains ....................................... 1-7
      Magnetic Field ............................... 1-12
      Lines of Force ............................... 1-13
   Magnetic Particle Test Principles ................ 1-18
      Magnetic Poles ............................... 1-18
      Formation of Magnetic Poles .................. 1-30
      Ferromagnetic Material ....................... 1-38
      Residual Magnetism ........................... 1-43
      Permeability ................................. 1-46
      Flux Density ................................. 1-49
      Retentivity .................................. 1-59
      Coercive Force ............................... 1-64
      Hysteresis Loop .............................. 1-72
   Review ........................................... 1-80

Chapter 2 - Producing Magnetic Fields ............... 2-1

   Magnetizing current .............................. 2-1
      Right-Hand Rule .............................. 2-2
   Circular Magnetic Fields ......................... 2-6
      In Ferromagnetic Materials ................... 2-6
      In Rectangularly-Shaped Bars ................ 2-12
      In Nonmagnetic Materials .................... 2-13
      Poles Formed by Discontinuities ............. 2-15
      Producing a Circular Field .................. 2-22
      Distribution of the Circular Field .......... 2-39
      Use of a Central Conductor .................. 2-41

| | |
|---|---|
| Longitudinal Magnetic Fields | 2-52 |
|     Poles Formed by Discontinuities | 2-52 |
|     Magnetization by Coil | 2-56 |
|     Field Strength in a Coil | 2-59 |
|     Effective Field | 2-69 |
| Magnetization by Cable | 2-71 |
| Use of Prods | 2-74 |
| Use of a Yoke | 2-81 |
| Review | 2-82 |

## Chapter 3 - Magnetizing Currents .... 3-1

| | |
|---|---|
| Alternating Current | 3-1 |
| Direct Current | 3-3 |
|     Flux Distribution (AC) | 3-12 |
|     Flux Distribution (DC) | 3-16 |
| Current Requirements | 3-21 |
|     Circular Magnetization | 3-21 |
|     With Prods | 3-35 |
|     Longitudinal Magnetization | 3-37 |
| Review | 3-49 |

## Chapter 4 - Materials and Sensitivity .... 4-1

| | |
|---|---|
| Particles | 4-1 |
|     Wet Bath | 4-5 |
|     Dry Particles | 4-12 |
|     Sensitivity of Mediums | 4-14 |
| Review | 4-28 |

| | |
|---|---|
| Self Test | A-1 |
| Glossary | B-1 |
| Measurement Conversion Charts | C-1 |

# INSTRUCTIONS

The pages in this book should **not** be read consecutively as in a conventional book. You will be guided through the book as you read. For example, after reading page 3-12, you may find an instruction similar to one of the following at the bottom of the page:

- Turn to the next page.

- Turn ahead to page 3-15.

- Turn back to page 3-8.

On many of the pages you will be faced with a choice. For instance, you may find a statement or question at the bottom of the page together with two or more possible answers. Each answer will indicate a page number. You should choose the answer you think is most correct and turn to the indicated page. If you happen to select an incorrect answer, continue to read, as the page will provide supplemental information to help you understand the concept.

We know that sometimes the information in this self-study format may seem oversimplified or repetitious. Bear with us; the reinforcement of basic Level I concepts is essential if you expect to retain the knowledge and apply it to Level II training or on-the-job NDT applications.

Do not rush through the volumes. Take whatever time you need to make sure you have a clear understanding of the material. Depending on your background knowledge, reading speed, etc., the time needed to complete this book may vary from **15 to 20 hours** or more. As you will soon see, this self-study handbook is easy to use—just follow instructions.

TURN TO THE BEGINNING OF CHAPTER 1.

# CHAPTER 1

## MAGNETISM

### Theory

Magnetic particle testing is a nondestructive test method for detecting discontinuities at or near the surface of materials that can be strongly magnetized. To more fully appreciate the details of how the magnetic particle test functions, let us first study elementary magnetism and magnetic field theory.

A *magnet* is a material that has the ability to attract iron or steel (and some other metallic materials). Lodestone (magnetite) is naturally magnetic. Other materials can become magnets artificially. When any material is magnetized it has a magnetic field that will attract certain metals and other magnetic fields.

Believe it or not, all materials can be classified as magnetic. That's right! All materials are *diamagnetic* materials yet purely diamagnetic materials do not make good magnets because their atomic arrangements won't allow it. *Paramagnetic* materials (such as aluminum) will atomically align themselves with a nearby magnetic field but will not themselves provide the magnetism needed for magnetic particle testing.

*Ferromagnetic* materials such as iron and cobalt can easily become strongly magnetized. Ferromagnetic materials are generally called "magnetic materials" or magnets.

Turn to the next page.

Let's consider the ordinary compass which is simply a magnetized (ferromagnetic) steel pointer. When the pointer is allowed to rotate freely, it will always point in the same direction—north. This is because the pointer is attracted by the Earth's natural magnetic field.

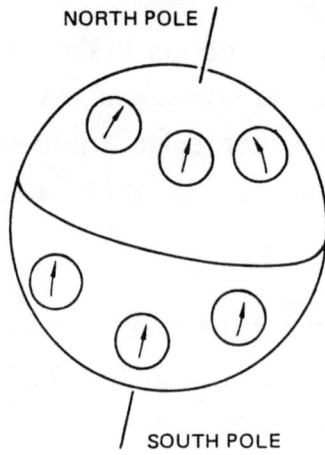

Observe that the compass needle points to the Earth's magnetic north pole from any point on the Earth's surface. (The Earth's magnetic north pole and geographic North Pole are not the same, though we'll not concern ourselves with this point.) Every magnet, like the Earth, has a north and south pole.

Turn to the next page.

Now, consider the Earth as a giant magnet and you can see why the magnetic compass acted as it did.

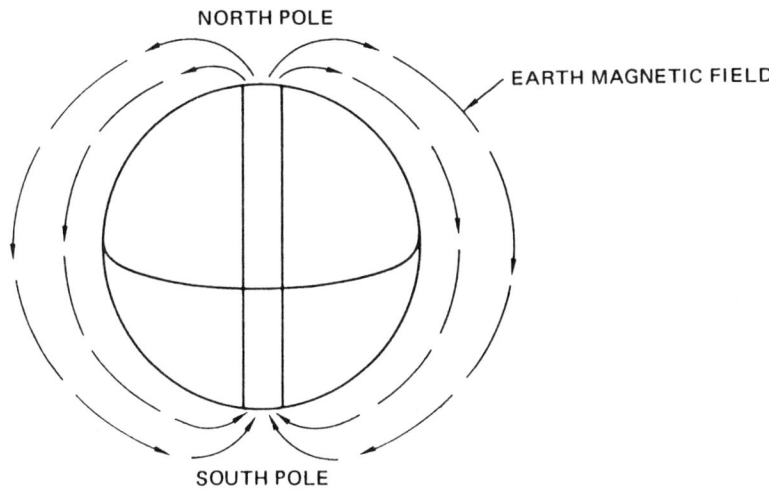

The magnetic compass acted as it did because of the Earth's *magnetic field*. The compass needle, being magnetized, was aligned with the Earth's magnetic field that enters and/or exits the Earth at it's magnetic north and magnetic south poles. This brings up the rule of magnetic attraction and repulsion for magnets. That is:

**LIKE MAGNETIC POLES REPEL AND UNLIKE MAGNETIC POLES ATTRACT ONE ANOTHER.**

Turn to the next page.

Using this rule, we know that if two magnets are placed so that a south pole of one is placed close to the north pole of the other, the magnets will be attracted to each other. What would happen if bar magnets are placed like this?

[ N    S ]         [ S    N ]

**Select the correct statement and turn to the page indicated.**

**The bar magnets would attract each other** ............ **Page 1-6**
**The bar magnets would repel each other** ............ **Page 1-8**

From page 1-8

Not exactly. A south pole will repel another south pole. They will tend to push away from one another because they are *like* poles.

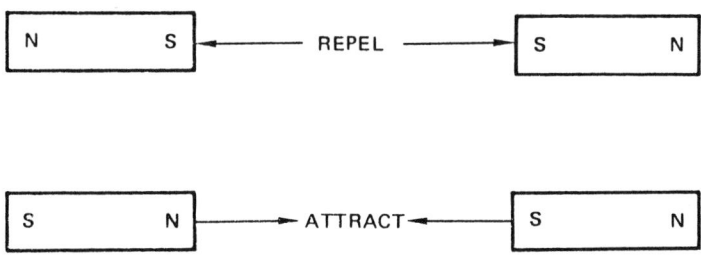

Let's look at the rule again.

**LIKE MAGNETIC POLES REPEL EACH OTHER.**
**UNLIKE MAGNETIC POLES ATTRACT EACH OTHER.**

Return to page 1-8 and select another alternative.

From page 1-4

You selected "The bar magnets would attract each other." Perhaps we've confused you by not telling you what we mean by "like poles." Consider a compass needle. The red or white, and normally luminous, end of the compass needle points to the Earth's magnetic north pole. The other end of the needle would then point to magnetic south. Here are two compasses held apart from one another.

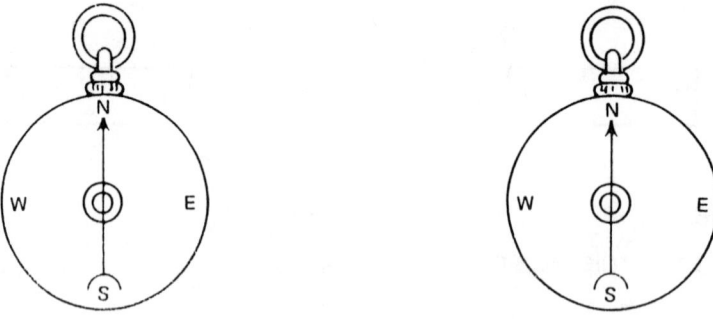

If we bring the two compasses close together, the needles will no longer point to the Earth's north pole. The north pole of one needle will attract the south pole of the other like this.

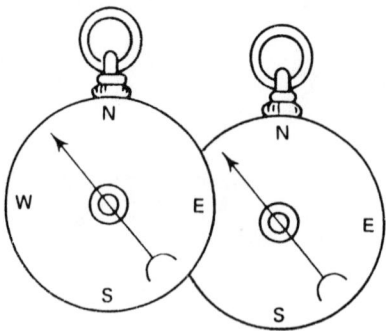

Here you can see the *unlike* (north and south) poles are attracting one another.

Turn ahead to page 1-8.

From page 1-8

Right again. The two bar magnets were attracted to each other because *unlike poles attract each other*. A south pole of one magnet was attracted by the north pole of the other magnet.

Now let's take a deeper look at the reasons why magnets act as they do.

Just as the Earth itself is a magnet having a north and south pole, every material (or element) is also one type of magnet or another having a north and south pole. The atoms in a material act collectively to set up an individual magnetic field or force.

Even in an unmagnetized piece of iron, "atomic magnets" exist and are arranged in a haphazard fashion like this. Let's call these "atomic magnets" *domains*.

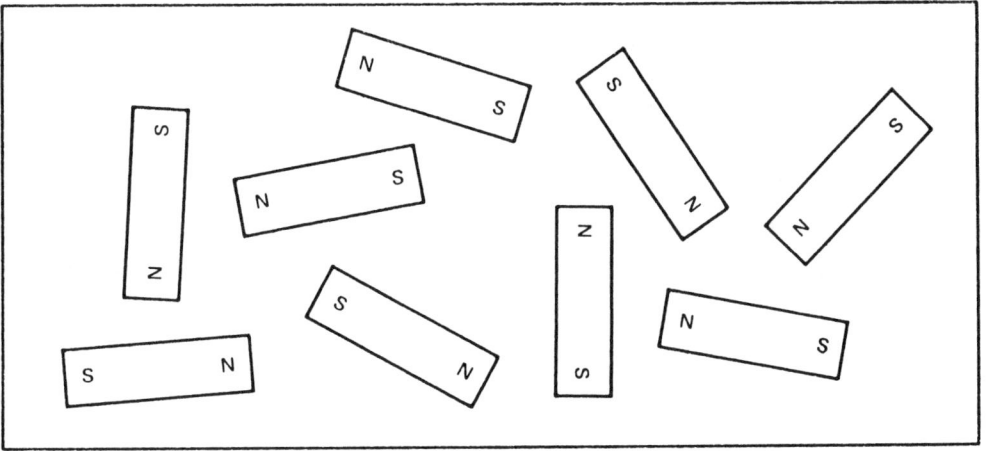

Now if all the domains in the above piece of iron were lined up in an organized manner, the piece of iron would be at its highest magnetic potential or strength.

Turn ahead to page 1-9.

From page 1-4    1-8

Right. If the bar magnets were placed so that their south poles were close together, they would repel each other. Here is the rule again.

**LIKE MAGNETIC POLES REPEL EACH OTHER.**
**UNLIKE MAGNETIC POLES ATTRACT EACH OTHER.**

Here are two bar magnets without their poles identified.

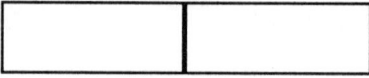

You can see that these magnets have attracted each other to the point of contact.

**Which of the following combinations of poles must exist for the bar magnets to attract each other?**

Two south poles . . . . . . . . . . . . . . . . . . . . . . . . . . . . . . . . **Page 1-5**
A south and a north pole . . . . . . . . . . . . . . . . . . . . . . . . . **Page 1-7**
Two north poles . . . . . . . . . . . . . . . . . . . . . . . . . . . . . . . **Page 1-10**

From page 1-7  1-9

Now, in a manner which will be discussed later, let's magnetize this same piece of iron. When the iron is magnetized, each "atomic magnet" of the iron is magnetized so that all north poles are oriented in the same direction and all of the south poles are oriented in the same direction.

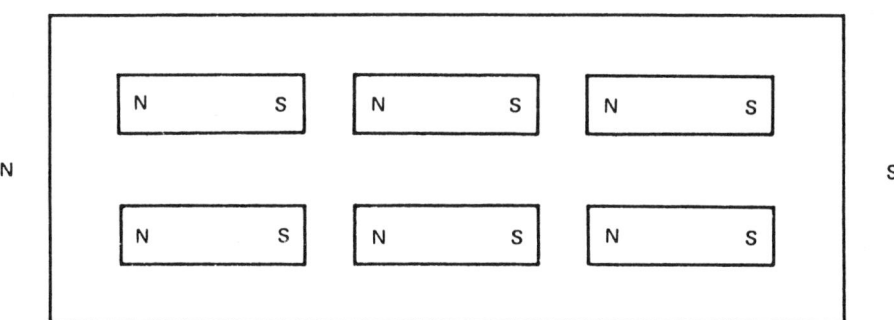

With all of the individual domains magnetized like this, the piece of iron then has a north and south pole.

**Since the iron is now magnetized so that the north pole of one domain is facing the south pole of the next, what rule of magnetism is taking place within the bar?**

Unlike magnetic poles repel each other ............. Page 1-11
Unlike magnetic poles attract each other ............ Page 1-13

No. A north pole will repel another north pole. They will tend to push away from one another because they are "like poles."

$$\boxed{S \quad N} \qquad \boxed{N \quad S}$$

Here is the rule again.

**LIKE MAGNETIC POLES REPEL EACH OTHER.**
**UNLIKE MAGNETIC POLES ATTRACT EACH OTHER.**

Turn back to page 1-8 and select another answer.

From page 1-9                                                           1-11

You say that unlike magnetic poles repel each other. Look at the bar again.

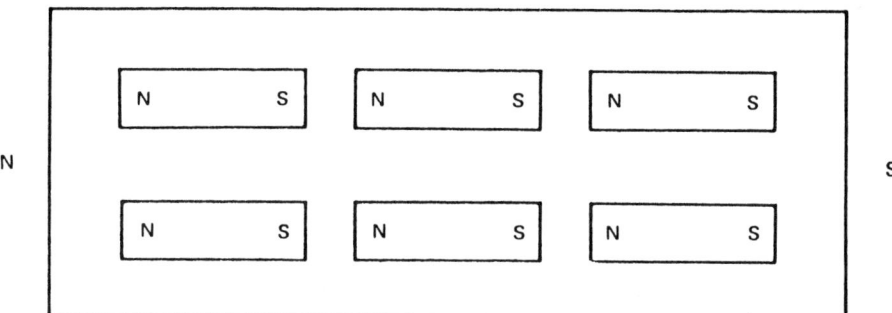

In this case, we have the south pole of one domain facing the north pole of another. When we have a situation like this, the poles attract each other.

**UNLIKE MAGNETIC POLES ATTRACT EACH OTHER.**

Turn ahead to page 1-13.

Right. Lines of force around a magnet follow a path to an opposite pole.

The space around a magnet in which the lines of force act is called the *magnetic field*.

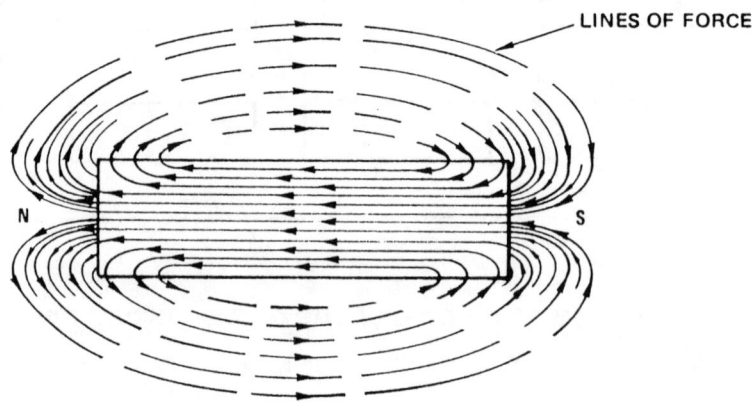

There are many lines of force surrounding a magnet. All of the lines of force make up the magnetic field. The field is most dense at the ends of the magnet where the field is strongest.

**Based on the above, which of the following statements is correct?**

**All magnetic lines of force around a magnet are contained within the magnetic field ........................... Page 1-15**
**Magnetic lines of force around a magnet are found both inside and outside the magnetic field ............... Page 1-17**

Good for you. That's right. Unlike poles attract each other. Since a north pole was facing a south pole, both poles were attracted to each other.

With all of the domains lined up this way, the magnetic bar develops a total force equal to the sum of all the forces of all the domains. This is what we have now.

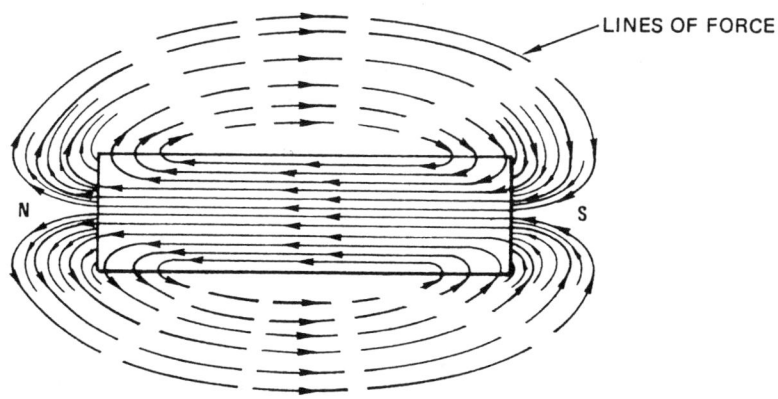

Shown are magnetic lines of force which surround every magnet. These lines of force have a definite direction. They leave their north pole and re-enter their south pole and continue on their way through the magnet from the south pole to the north pole.

Turn to the next page.

From page 1-13

Magnetic lines of force are continuous and always form a closed loop or circuit. The individual lines of force do not cross or merge with other lines of force.

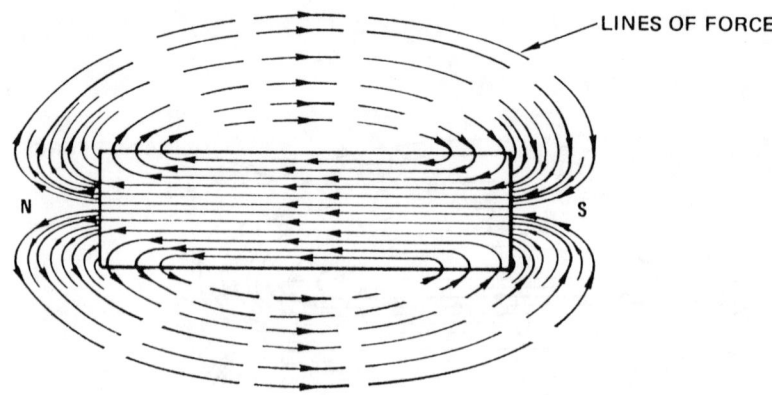

**Considering the direction the lines of force take around a magnet, which of the following statements is true?**

**Lines of force follow a path to an opposite pole** ...... **Page 1-12**
**Lines of force follow a path to a like pole** ............ **Page 1-16**

That's right. All magnetic lines of force around a magnet are contained within the magnetic field. Or better yet, the magnetic field consists of *all* the lines of force in and around the magnet.

Now let us look at our bar magnet without the external lines of force.

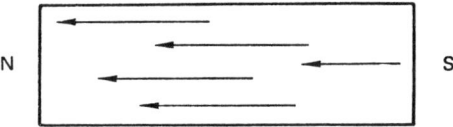

Here you can see the lines of force within the magnet flowing from the south pole to the north pole. Let's break the bar magnet into several pieces.

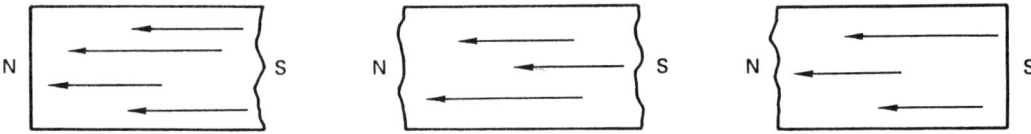

When the magnet is broken into several pieces, each piece becomes a complete bar magnet within itself with a north and south pole and lines of force. If we continue to break the bar into more pieces, each piece will have a north pole and a south pole.

Turn ahead to page 1-18.

You feel that lines of force follow a path to like poles. Let's look at this from a different angle. Here is a magnet with the lines of force going from the south pole to the north pole within the bar magnet.

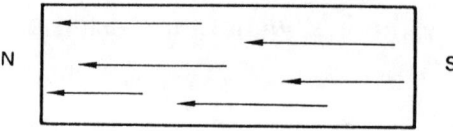

Now, let's add the external lines of force.

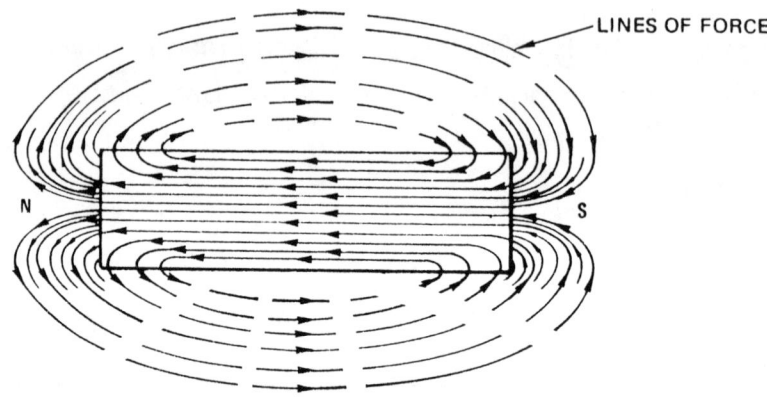

As you can see, in both cases, the lines of force *always* follow a path to an opposite pole. Note that the lines of force leave and enter the magnet at or near the ends of the magnet. They *do not* leave or return at the center of the magnet.

Turn back to page 1-12.

From page 1-12   1-17

You weren't thinking on this one. It is impossible to have any magnetic lines of force outside the magnetic field since the magnetic lines of force *are* the magnetic field.

**ALL MAGNETIC LINES OF FORCE AROUND A MAGNET ARE CONTAINED WITHIN THE MAGNETIC FIELD.**

Turn back to page 1-15.

## Magnetic Particle Test Principles

In discussing the Theory of Magnetism, we found that every magnet has a north and a south pole. We found that every magnet also has a magnetic field comprised of magnetic lines of force.

Now we are going to discuss these terms as they apply to the principles of magnetic particle testing.

The magnetic field is the force that attracts other magnetizable materials to the magnetic poles. Another name for these lines of force is *magnetic flux lines*. Often the term "*magnetic flux*" is used to denote the entire magnetic field so we may say that the magnetic flux is the force that attracts magnetizable material to the magnetic poles.

First, let us study the nature of the *magnetic flux* in long-lived (or "permanent") magnets of different shapes.

A common shape for a magnet is the horseshoe which is known as a horseshoe magnet.

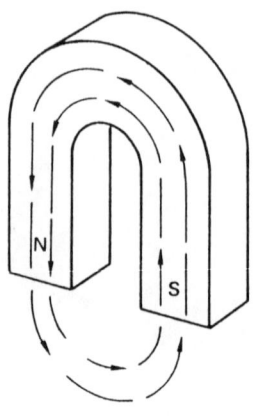

Turn to the next page.

From page 1-18                                                          1-19

In the horseshoe magnet, the magnetic flux or lines of force will enter or leave the magnet at the poles. The horseshoe magnet will attract other magnetizable material only where the lines of force *leave or enter* the magnet.

**If we were to dip the horseshoe magnet into a bucket of iron filings, where could we expect magnetic flux to attract the filings?**

**At the north pole** .................................. **Page 1-20**
**Anywhere on the magnet** ....................... **Page 1-22**
**At the north and south poles** ................. **Page 1-25**

From page 1-19

You are half right. Iron filings would be attracted to the north pole. That is because the lines of force *leave* the magnet at the north pole.

In the horseshoe magnet, the magnetic lines of force or flux will enter or leave the magnet at the poles.

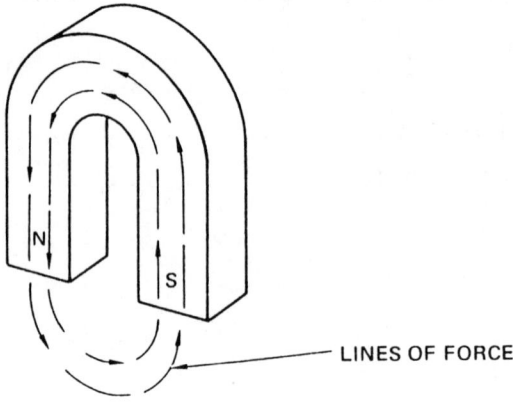
LINES OF FORCE

The horseshoe magnet will attract other magnetizable materials only where the lines of force *leave* or *enter* the magnet.

Turn back to page 1-19 and try one of the other answers.

From Page 1-25                                                                    1-21

That's right.  If the steel bar were placed across only the north pole, it would be attracted to the magnet because the magnetic flux lines *leave* the magnet at the north pole.  If the steel bar had been placed over the south pole it would also have been attracted since that is where the lines of force *enter* the magnet.

Actually, we can say that magnetizable iron and steel will be attracted to the poles of a magnet.

Suppose we bend the horseshoe magnet so the north and south poles are close together.

In the circular, "doughnut"-shaped magnet, the lines of force flow across the gap from the north pole to the south pole.

**Where would iron filings be attracted to this circular magnet?**

**Nowhere on the magnet** ......................... **Page 1-23**
**At the poles** .................................... **Page 1-26**

From page 1-19  1-22

You selected "anywhere on the magnet." You have missed the key words. Magnetizable material will be attracted to the horseshoe magnet *only where the lines of force leave or enter the magnet*.

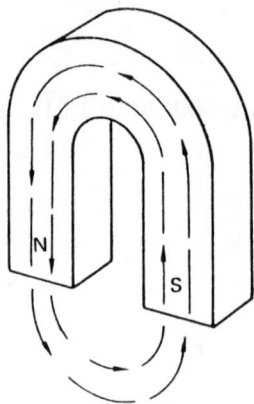

In the metal portion of this magnet, the magnetic field is contained entirely within the metal. The lines of force go from the south pole to the north pole and do not enter or leave the magnet between these end points. At the north pole, the lines of force leave the magnet. These lines of force are attracted by the south pole where they re-enter the magnet. Since magnetizable materials, like iron filings, will be attracted to the magnet where the lines of force leave or enter the magnet, iron filings will be attracted only to the north and south poles of the magnet.

Turn back to page 1-19 and choose another answer.

From page 1-21                                                                                   1-23

You think that iron filings would not be attracted anywhere on the magnet? Most of the circular magnet will not attract iron filings. But we still have the poles, created by the gap, where the lines of force enter and leave the magnet.

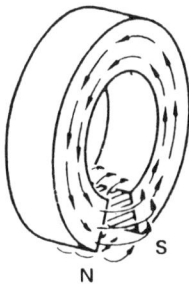

Although we did change the shape of the magnet, we did not change the fact that there are two magnetic poles. As you can see in the illustration above, the lines of force are leaving the magnet at the north pole and re-entering the magnet at the south pole. Iron filings would be attracted to these poles.

Turn ahead to page 1-26.

From page 1-25

You think that the bar would not be attracted to the north pole. You must have forgotten that magnetic materials will be attracted to the magnet at the points where the lines of force *enter* or *leave* the magnet. Here, you can see the lines of force leaving the magnet at the north pole and entering the magnet at the south pole.

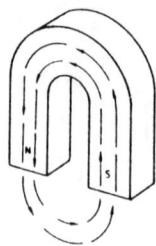

The steel bar will be attracted to the north pole.

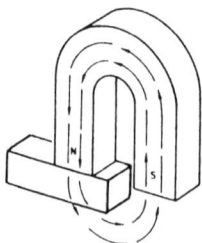

The steel bar will also be attracted to the south pole.

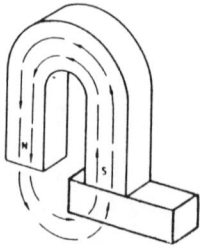

The steel bar will be attracted to either pole because these are the places where the lines of force *enter* and *leave* the magnet. Right?

Turn back to page 1-21.

From page 1-19                                                                  1-25

Right. If we dip the horseshoe magnet into a bucket of iron filings, the magnetic flux would attract iron filings to both the north and south poles. Why? Because a horseshoe magnet will attract another magnetizable material ONLY where the MAGNETIC FLUX (lines of force) leave or enter the magnet. In fact, we can identify an object as being a magnet if it will attract other magnetizable materials, such as iron filings, when placed close together.

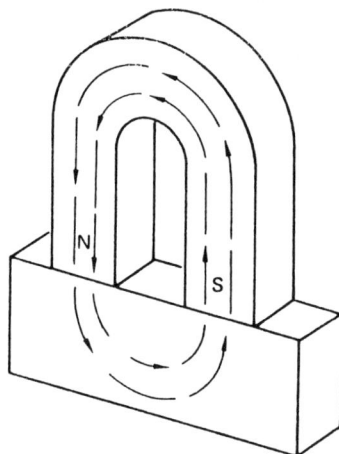

Here, a magnetizable steel bar has been placed across the poles of the magnet. It is held in place by the attracting force of the magnetic flux. The magnetic flux lines flow from the north pole of the magnet through the steel bar to the south pole of the magnet.

**If the steel bar were placed across only the north pole, would it be attracted to the horseshoe magnet?**

**Yes** .................................................... **Page 1-21**
**No** ..................................................... **Page 1-24**

From page 1-21

Good for you. That's right. Iron filings would be attracted at the poles.

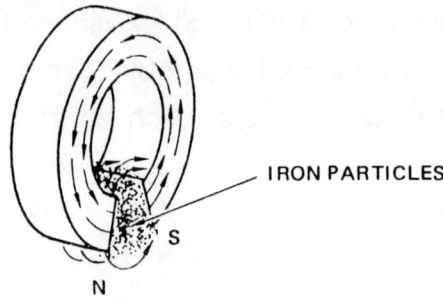

Here you can see the iron filings (or particles) clinging to the poles and bridging the gap between the poles.

Turn to the next page.

Now, let's make a complete circle out of the magnet and see what happens.

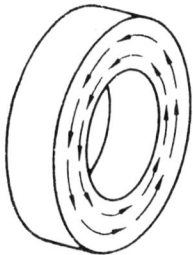

Here is a circular or "doughnut" magnet without any poles. The lines of force or flux are contained entirely within the circle.

**If we dust particles of iron on this doughnut magnet, where would they be attracted to the magnet?**

**Iron particles would be attracted to all points on the magnet** .................................................. **Page 1-28**
**Iron particles would NOT be attracted to the magnet** ... **Page 1-30**

Watch out! Did you see any magnetic *poles* on that doughnut magnet?

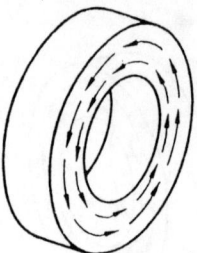

Remember, iron particles will be attracted *only* where lines of force or flux enter or leave the magnet. Since there are no magnetic poles, there will be no place for the magnetic lines of force or flux to leave or enter the magnet.

Turn ahead to page 1-30.

You think that iron particles would *not* be attracted at the flux leakage? Let's explain a little more fully what is happening. Here is an enlarged view of the crack in the doughnut magnet.

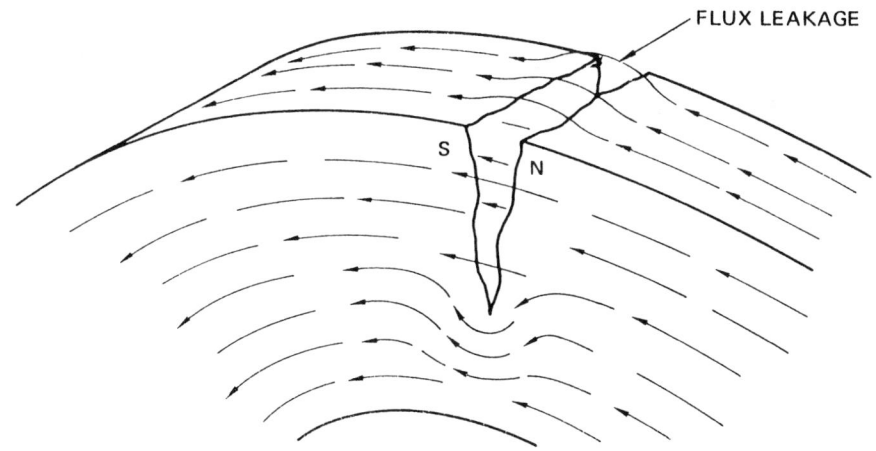

This crack runs crosswise or perpendicular to the lines of force. At the crack, the lines of force pass through the crack causing a north and south pole. Some of the lines of force are redirected and pass over and under the crack. They do this because they are following the path of least resistance. The lines of force that pass through and over the crack are known as *flux leakage*. This *flux leakage* is what attracts iron particles.

Turn ahead to page 1-33.

From page 1-27

Excellent. That's right. Iron particles would *not* be attracted to the magnet at all. Since the circle magnet has all of the lines of force contained within the magnet, there is no place where the lines of force or flux can enter or leave. In other words, there are no poles.

Let us now take a look at a doughnut magnet with a crack in the outer surface and see what happens.

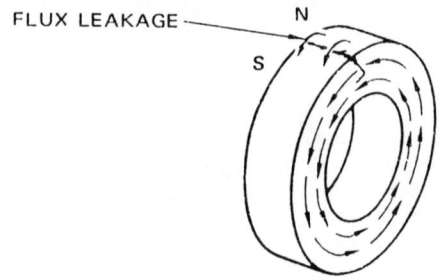

Any crack in the magnet will disrupt the internal flow of the lines of force. Some of the lines of force will be driven out of the magnet. This creates an external magnetic field and establishes a north and south pole. The lines of force that are redirected or driven out of the magnet as a result of the crack are known as *flux leakage*. The effect is similar to the breaking of the bar magnet into pieces we discussed earlier.

**Since the crack in the doughnut or circular magnet has created a north and south pole, what would you expect to occur where the flux leakage is located?**

**Iron particles would not be attracted** . . . . . . . . . . . . . . . . . Page 1-29
**Iron particles would be attracted** . . . . . . . . . . . . . . . . . . . Page 1-33

From page 1-33                                                              1-31

The bar magnet has the same characteristics as the horseshoe magnet. The lines of force or flux flow from the south to the north pole within the magnet. Iron particles will be attracted only to the poles where the lines of force or flux leave and enter the magnet.

A crack in a bar magnet will also cause *flux leakage*.

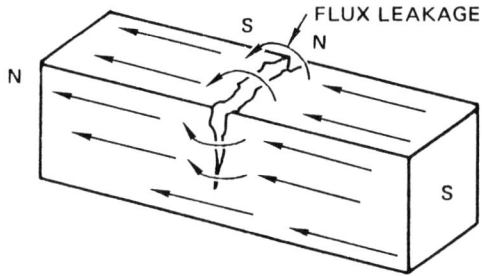

The lines of force at the bottom of the crack tend to follow the line of least resistance and remain in the magnet. The lines of force passing through the area of the crack tend to be forced to the surface. Some of the lines of force bridge the gap and pass through the crack, while others are forced to the surface where they pass over the crack. Those lines of force that pass through and over the crack cause flux leakage and establish north and south poles in the vicinity of the crack.

**Do you think iron particles would be attracted at the flux leakage created by the crack?**

**Yes** .................................................. Page 1-32
**No** ................................................... Page 1-34

From page 1-31

Excellent. Of course iron particles would be attracted at the flux leakage created at the crack.

If we had a bar magnet with a slot cut into it like this one, we will have *flux leakage*.

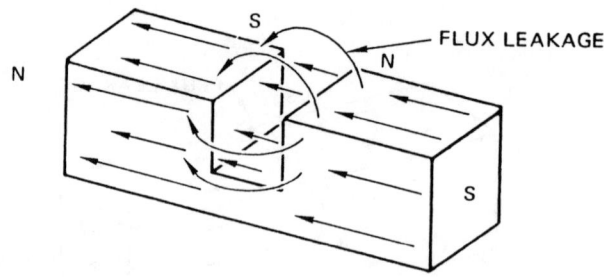

Here you can see the magnetic poles and flux leakage created by the slot. The lines of force in the vicinity of the slot tend to be forced toward the surface. Some of the lines of force jump through the slot, while others are forced to the surface where they pass over the slot.

**If we add two more slots to the above magnet, do you think each would create flux leakage?**

No ............................................... Page 1-35
Yes .............................................. Page 1-36

Correct. You would expect iron particles to be attracted at the crack where flux leakage is located. The iron particles would be attracted to the poles created by the crack. Here is the cracked doughnut magnet again.

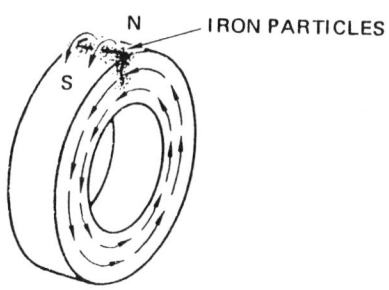

Here you can see that the iron particles have been attracted by the *flux leakage* created by the crack.

Now let us go back to the horseshoe magnet. If we straighten the horseshoe magnet, we have our bar magnet.

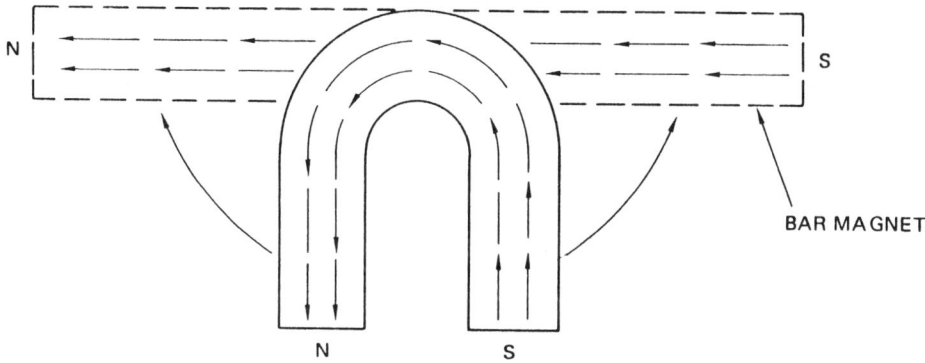

Turn back to page 1-31.

You have missed the point. Let's look at that diagram again.

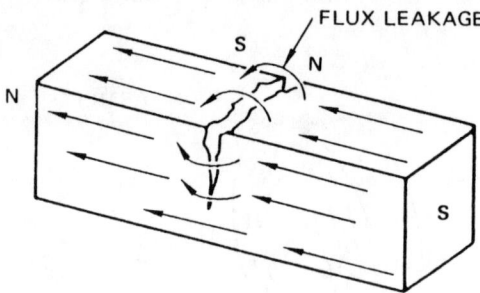

Review these facts:

- We have a magnetized bar (bar magnet).

- The lines of force pass through the bar.

- The lines of force are interrupted by the crack, causing *flux leakage*.

- Magnetic poles are formed at the crack.

- Iron particles will be attracted by *flux leakage* at the magnetic poles formed at the crack.

Turn back to page 1-32.

Evidently we haven't made the point clear. Let's go back to the example of the broken bar magnet.

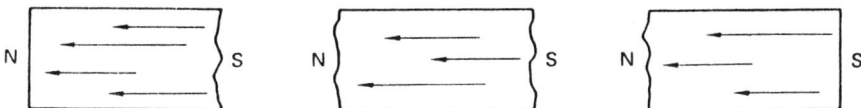

When the bar magnet is broken into several pieces, each piece becomes a complete magnet within itself with a north and south pole and lines of force. The same thing happens to the bar magnet if we cut slots in it.

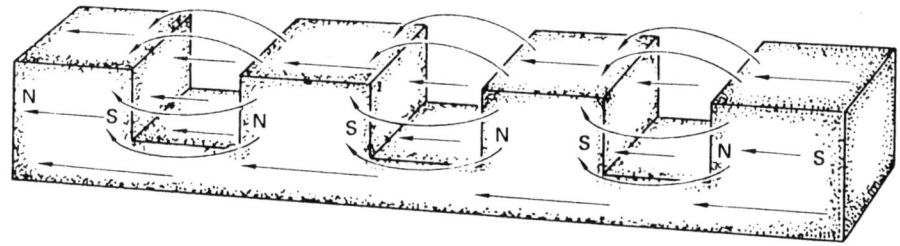

If we cut three slots in the bar, a north and south pole will be created at each slot. We will also have flux leakage at each of the slots as shown above.

Turn to the next page.

Yes, each slot that we put into that magnet would create flux leakage. Iron particles would be attracted by the flux leakage.

On any magnet, magnetizable materials, like iron, will be attracted to the poles of the magnet. If the magnet has all of the lines of force contained within the magnet as with the doughnut magnet, there would be no poles. Therefore, iron particles would not be attracted.

Now let's look at a magnet with a shallow surface irregularity, such as a bowed or cupped surface.

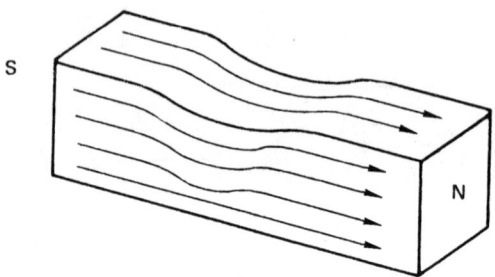

In the area of the shallow, cupped surface above, the lines of force stay within the magnet. The lines of force tend to follow the path of least resistance, which is to stay within the magnet. As a result, no magnetic poles with flux leakage are created.

**Would iron particles be attracted to the shallow, bowed surface above?**

**No** . . . . . . . . . . . . . . . . . . . . . . . . . . . . . . . . . . . . . . . . . **Page 1-37**
**Yes** . . . . . . . . . . . . . . . . . . . . . . . . . . . . . . . . . . . . . . . . **Page 1-39**

From page 1-36 1-37

Of course not. There were no poles with flux leakage to attract iron particles. The lines of force followed the path of least resistance which was to follow the metal in the shallow, cupped surface. Similarly, a shallow scratch would cause only a weak, if any, leakage field.

Here's another magnet with a crack located below the surface.

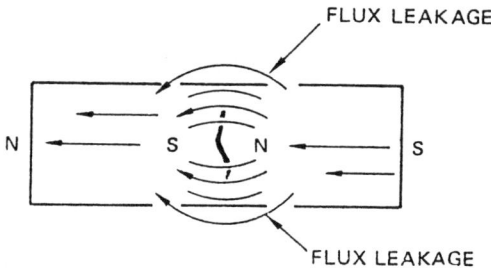

With this subsurface crack, you can see that some of the lines of force pass above and below the crack. Some of the lines of force pass through the crack and some are forced out at the surface, creating a localized flux leakage.

**Do you think that iron particles would be attracted to the flux leakage caused by the subsurface crack?**

**Yes** .................................... **Page 1-38**
**No** ..................................... **Page 1-41**

From page 1-37                                                                 1-38

That's right. Iron particles would be attracted to the flux leakage caused by the subsurface crack.

Magnetic particle test principles depend upon establishing a magnetic field within a test specimen. Therefore, the specimen to be inspected must be made of materials which can be strongly magnetized. *Ferrous* materials are most strongly affected by magnetism.

By definition, **ferrous** means **"pertaining to or derived from iron."** Since iron can be easily magnetized, it is called *ferromagnetic*. Iron, most steels, nickel, cobalt, and many of their alloys are ferromagnetic materials.

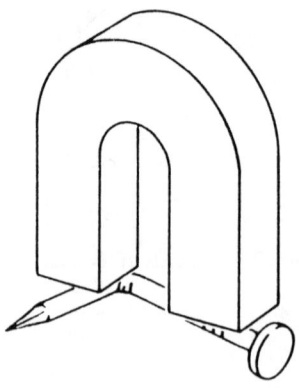

If a nail can be picked up by a horseshoe magnet, what kind of material would you say the nail is made of?

A nonferromagnetic material . . . . . . . . . . . . . . . . . . . . . . . Page 1-40
A ferromagnetic material . . . . . . . . . . . . . . . . . . . . . . . . . Page 1-42

We caught you napping. Iron particles would NOT be attracted to that shallow, rounded surface. Because the lines of force remained in the metal, *no flux leakage was created*.

Remember, iron particles will only be attracted at points where the lines of force *leave* and *enter* the magnet. In other words, iron particles will only be attracted to flux leakage. In the example, there was no flux leakage at the shallow, rounded surface, so iron particles would not be attracted.

Turn back to page 1-37.

Oooops! You've got it backwards. *Ferromagnetic* means **"easily magnetized—attracted by magnets."**

**Nonferromagnetic** means **"not easily magnetized—not attracted by magnets."**

The nail is strongly attracted by the magnet; therefore, it is a ferromagnetic material.

Turn ahead to page 1-42.

From page 1-37                                                          1-41

You have forgotten one thing: WHEREVER LINES OF FORCE ENTER OR LEAVE THE METAL, POLES WILL BE FORMED AND IRON PARTICLES WILL BE ATTRACTED TO THE FLUX LEAKAGE.

*Flux leakage* will also be formed whenever a crack below the surface causes the lines of force to leave the metal.

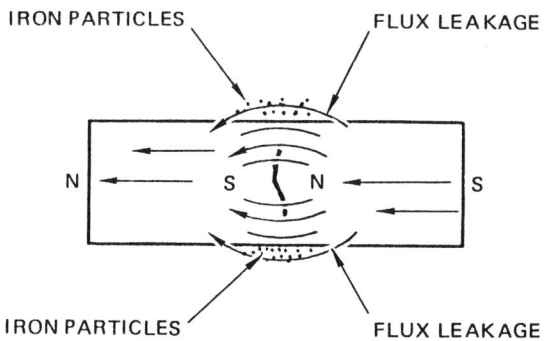

Notice that the spot where lines of force leave the metal is not as clearly defined as it would be if there was a crack in the surface. So, *iron particles would be attracted to the flux leakage caused by the subsurface crack*.

Turn back to page 1-38.

You bet. The nail would have to be ferromagnetic material in order to be picked up by the magnet. Ferromagnetic materials are a special group of paramagnetic materials which are strongly attracted by any magnetic field and can, in turn, be magnetized. However, most paramagnetic materials may only temporarily become very slightly magnetized. Examples of these materials include aluminum, brass, copper, magnesium, bronze, lead, titanium, and some stainless steels.

**If a piece of wire is NOT attracted to a horseshoe magnet, you would know that the wire is made of which kind of material?**

**Ferromagnetic material** . . . . . . . . . . . . . . . . . . . . . . . . . . . . **Page 1-44**
**Nonferromagnetic material** . . . . . . . . . . . . . . . . . . . . . . . . **Page 1-46**

From page 1-47

Right. Of course the horseshoe magnet is not highly permeable.

Electric current can and often is used to create a magnetic field in magnetic material. The magnetic field that remains in the magnetic material after the magnetizing current is shut off is called *residual magnetism*.

Illustrated below are both magnetically soft iron (top) and magnetically hard steel (bottom) materials. The left portion of the illustrations indicate the lines of force while the magnetizing current is applied. The right hand portion illustrates the remaining lines of force after the current is shut off. Note that soft iron is easy to magnetize and is *highly permeable*. Very hard ferromagnetic steel has *low permeability* and is hard to magnetize.

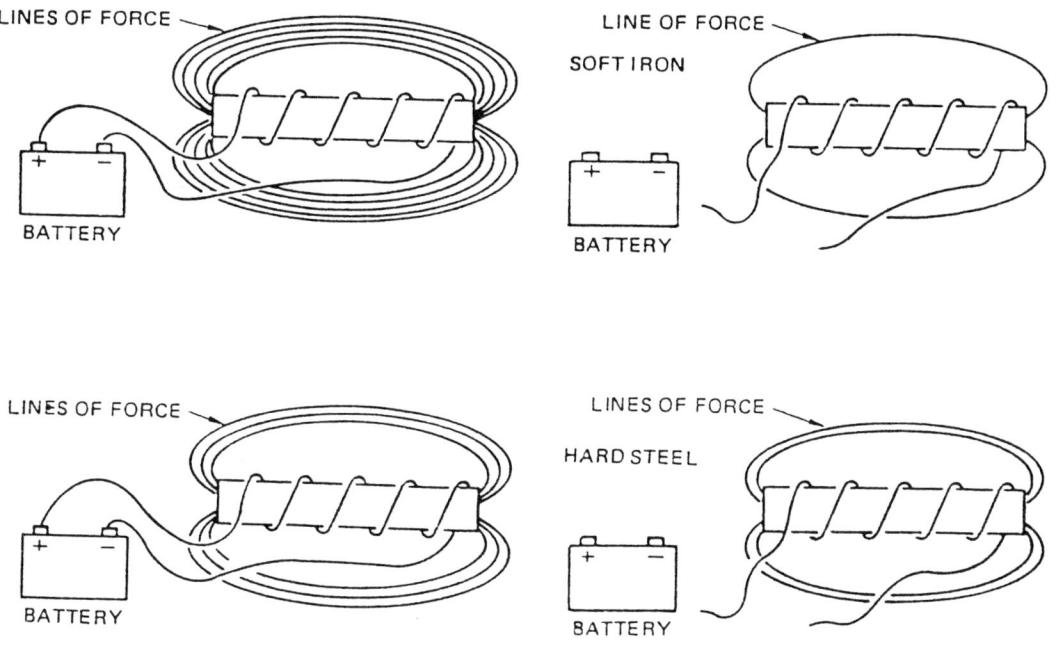

Turn ahead to page 1-48.

From page 1-42

Incorrect. A piece of wire that would not stick to a horseshoe magnet would not be ferromagnetic material.

Ferromagnetic materials will be attracted to a magnet. A nail is ferromagnetic, since it can be picked up with a horseshoe magnet. Remember, **ferrous** means **"of or pertaining to iron."** Iron will be attracted to a magnet.

A piece of copper wire cannot be picked up by a magnet, so the wire is nonferromagnetic. Any metal that is not strongly attracted to a magnet is nonferromagnetic.

Turn ahead to page 1-46.

From page 1-47

You think the horseshoe magnet is highly permeable. It is just the opposite. The word permeable may be causing the misunderstanding.

*Permeability* comes from the word "permeate" meaning to spread through. **Permeability**, as we are using it, means **"the ease with which the lines of force are able to pass through the metal."**

High permeability means that it is easy for the lines of force to spread through the metal.

Low permeability means that it is hard for the lines of force to spread through the material.

Remember:

**HIGH PERMEABILITY... EASY TO MAGNETIZE.**
**LOW PERMEABILITY... DIFFICULT TO MAGNETIZE.**

Ferromagnetic steel with a high carbon content is difficult to magnetize; therefore, it has low permeability. The horseshoe magnet has *low permeability*.

Turn back to page 1-43.

Right. The wire would be made of nonferromagnetic material.

All matter is subject to the influence of a magnetic field to some degree. In other words, they are *permeable* to some small degree. A few types of materials, such as bismuth, are repelled by a magnetic field. Only a few materials are strongly attracted by a magnet. These are a special group of materials that we will label as *ferromagnetic*. We are concerned only with materials such as iron, most steels, nickel, cobalt and their alloys. We will refer to them as magnetic materials from here on. All other materials are commonly referred to as *nonmagnetic* for purposes of magnetic particle testing.

**PERMEABILITY** is defined as:

**"THE EASE WITH WHICH MAGNETIC LINES OF FORCE ARE CARRIED OR PASS THROUGH MATERIALS."**

"Permeability" comes from the word "permeate" meaning "to spread through." As we are using it, permeability means the ease with which the lines of force pass through a material.

Turn to the next page.

Soft iron and iron with a low carbon content are very easy to magnetize and are *highly permeable*. These magnetic materials readily conduct the lines of force or flux.

Magnetic materials that are hard to magnetize have *low permeability*. Hardened ferromagnetic steel with a high carbon content is HARD to magnetize and has *low permeability*.

A horseshoe magnet is a typical permanent (long-lived) magnet and is made of very hard, high-carbon-content ferromagnetic steel.

**Would you say that the horseshoe magnet is highly permeable?**

No . . . . . . . . . . . . . . . . . . . . . . . . . . . . . . . . . . . . . . . . . Page 1-43
Yes . . . . . . . . . . . . . . . . . . . . . . . . . . . . . . . . . . . . . . . . Page 1-45

From page 1-43                                                              1-48

Although hard, ferromagnetic steel has low permeability and is difficult to magnetize, it will hold some of the magnetism after the magnetizing current is shut off. That is how a permanent magnet like the horseshoe magnet is made. Remember, the magnetism retained in a magnet is called residual magnetism.

**Which of the following types of materials do you think would retain the most residual magnetism?**

**Magnetic material with high permeability** ............ **Page 1-50**
**Magnetic material with low permeability** ............ **Page 1-53**

Right. By increasing the magnetizing force (electric current strength), we increase the number of lines of flux. *Flux density* would increase. Thus, we increase the strength of the magnetic field.

**Flux density** is defined as **"the number of lines of force per unit area."**

By placing a piece of magnetic material inside the coil, a magnetic field is induced into the material.

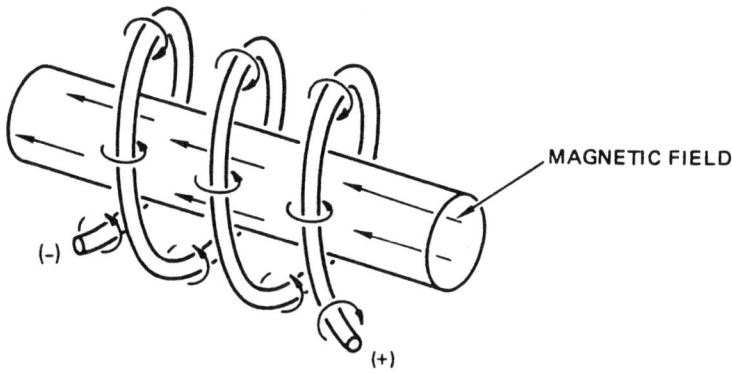

The maximum permeability of this particular material can be determined by increasing the magnetizing force (electric current strength) until the material reaches the point at which it can "hold" no more magnetism (its *saturation point*). Since different materials have different saturation points, we can determine the maximum permeability of each type of material.

Turn ahead to page 1-55.

From page 1-48                                                      1-50

You feel that magnetic material with high permeability would retain the most *residual magnetism*. That is incorrect and here is why.

Soft iron is easy to magnetize and is highly permeable. While these magnetic materials are highly permeable, they retain or hold very little of the residual magnetism after the magnetizing current is shut off.

Remember, **high permeability** means **low residual magnetism.**

Turn ahead to page 1-53.

From page 1-53

Very good. Magnetic materials with low permeability would have strong residual magnetism.

The permeability of a given material can be determined. As you will recall, electric current is used to create a magnetic field. A piece of copper wire wound into a coil will create a magnetic field when electric current is passed through the wire.

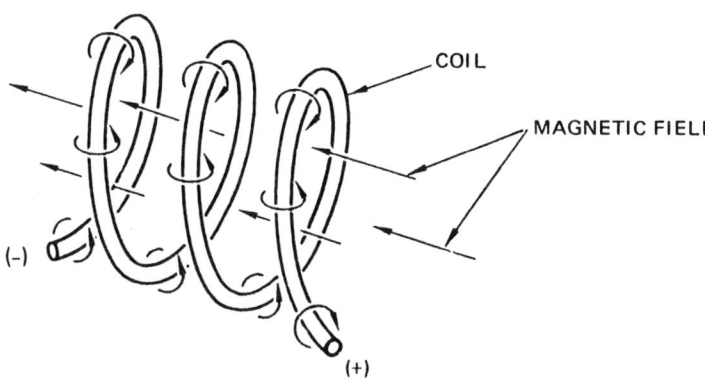

By varying the amount of electric current through the wire, we can vary the number of lines of force or flux within the coil. The total number of lines of flux is called *magnetic flux*.

**If we increase the current in the wire, what do you think would happen to the magnetic flux?**

**Magnetic flux would increase** ...................... **Page 1-49**
**Magnetic flux would decrease** ...................... **Page 1-52**

From page 1-51                                                    1-52

You selected "magnetic flux would decrease." Actually, it is just the opposite.

Without electric current flowing in the copper wire there is no magnetic field. When the electric current is turned on at a low current setting, the magnetic field is established. The total number of lines of force about the coil is called the magnetic flux. By increasing the electric current flowing in the wire, more lines of flux are formed. So you see, when we increase the amount of current in the wire of the coil, the magnetic flux will increase.

Turn back to page 1-49.

Correct. Magnetic material with low permeability would retain the most residual magnetism. Residual magnetism is the magnetic field retained in the material after the magnetizing current is shut off.

Residual magnetism is always less than the magnetic field which is present when the magnetizing current is on.

The amount of residual magnetism retained by a magnetic material will vary with the kind of material. For example, tool steel with a high carbon content will retain a stronger residual magnetic field than will steel with a low carbon content. A permanent magnet occurring in nature is the mineral *magnetite*. Other permanent magnets are made by magnetizing hardened steels or alloys such as *alnico*. Alnico is a steel alloy containing aluminum, nickel, cobalt and copper.

Soft magnetic material, such as iron and iron with a low carbon content, is very easily magnetized and is highly permeable. Unlike hard steel, soft iron will retain only a small amount of magnetism after the magnetizing current is removed. (Refer to the illustration on Page 1-43.) Soft iron retains very little residual magnetism.

**Magnetic materials with low permeability would have which of the following?**

**Strong residual magnetism** .......................... **Page 1-51**
**Weak residual magnetism** .......................... **Page 1-54**

From page 1-53

You think that material of low permeability would have weak residual magnetism. You have it backwards.

Remember:

**high permeability means easily magnetized . . . weak residual magnetism.**

**low permeability means difficult to magnetize . . . strong residual magnetism.**

Turn back to page 1-51.

From page 1-49

We can relate this "saturation" concept to the size of fuel tanks in automobiles. Some cars hold a 10-gallon (37.8 liter) maximum while others hold up to 22 gallons (83.3 liters). As another example, a sponge will only absorb so much fluid before it begins to drip. We would say that the sponge has become "saturated" once it begins to drip.

Similarly, each material will hold only a certain number of lines of flux or flux density regardless of how much we try to magnetize it.

**Therefore, we can say that each type of material has a point of:**

**minimum flux density** .............................. **Page 1-57**
**maximum flux density** .............................. **Page 1-58**

From page 1-58                                                          1-56

If we place a piece of unmagnetized or "virgin" ferromagnetic material in a coil and apply direct current through the coil—starting at zero amps and steadily increasing the current (magnetizing force)—we can plot the relationship between the magnetizing force, H, and the flux density, B, as the current increases. The resulting plot looks like this.

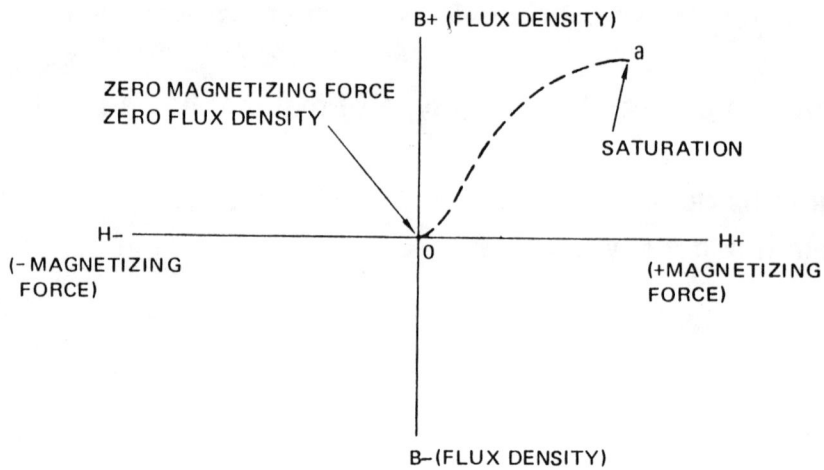

The plot tells us that as we increase the magnetizing force (move to the right along the H axis) the flux density increases (moves up along the B axis) until we reach the saturation point. Since the material was unmagnetized at the beginning, this curve is called the virgin curve for this material.

**Point "a" shows the point of saturation for the material. In other words, it shows:**

maximum flux density for the material . . . . . . . . . . . . . . . **Page 1-59**
maximum magnetizing force used . . . . . . . . . . . . . . . . . **Page 1-60**

From page 1-55

Well, each type of material has a minimum flux density all right—ZERO lines of flux. But we were talking about the saturation point of different materials . . . the point where each type of material cannot hold any more lines of flux. This is called the material's saturation point . . . its *maximum flux density.*

Turn to the next page.

That's right. Each type of material has a point of maximum flux density. At this point an increase in the magnetizing force will have no effect on flux density—the material is said to be saturated.

Every magnetic material has certain magnetic properties that are particular to that material. We would like to know what these are in order to determine how a material might respond to a magnetic particle test. Let's take a look at the response of a typical ferromagnetic material when we place it within a magnetizing force.

Turn back to page 1-56.

Very good. Point "a" on the virgin curve indicates the amount of magnetizing force used to obtain the maximum flux density for that material. Remember, along the dashed line the flux density increases as the magnetizing force is increased until it reaches a point at which any increase in the magnetizing force (H) does not increase the flux density (B). At this point (point "a") the material is saturated.

Now let's slowly reduce the magnetizing current; move left from saturation on the H axis. As we do so, we find that the path a - b does not retrace the path 0 - a.

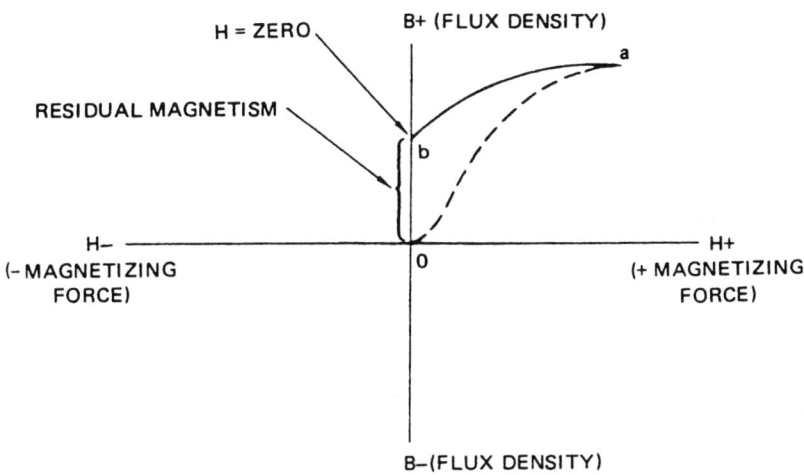

We find instead that when we have reduced the magnetizing force to zero (H=0) that we still have some flux density in the material. The flux density remaining is the *residual magnetism, remanence, or retentivity* and is shown by the distance 0 - b. This is what happens the first time any article is magnetized with direct current.

Turn ahead to page 1-61.

From page 1-56

You selected "maximum magnetizing force used." The chart didn't show that and here is the reason why.

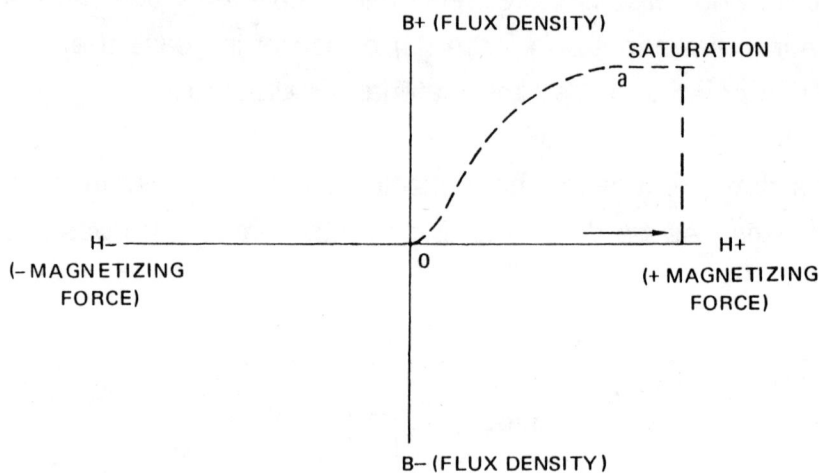

Actually, a considerable amount of *excess* electric current (magnetizing force) may have been used. Notice that the arrow points to the maximum magnetizing force which *could have been used* in this case. It is considerably greater than that needed to obtain the maximum flux density. So you see, the chart shows the maximum flux density for the material. The magnetizing force could have been any value above that needed to saturate the material.

Turn back to page 1-59.

From page 1-59

The ability of the material to retain a certain amount of residual magnetism is called *retentivity* or *remanence*.

**Which of the following do you think would have the greatest retentivity?**

**A material of high permeability** .................... **Page 1-62**
**A material of low permeability** .................... **Page 1-64**

From page 1-61

You think material of high permeability would have the greatest retentivity. Don't let that word "retentivity" throw you. Let's define it right here.

**Retentivity or remanence** is defined as:

**"THE ABILITY OF A MATERIAL TO RETAIN A PORTION OF THE MAGNETIC FIELD SET UP IN IT AFTER THE MAGNETIZING FORCE HAS BEEN REMOVED."**

Now, material of high permeability retains only a small amount of residual magnetism after the magnetizing force is removed. On the other hand, it is very easily magnetized. Soft iron and iron with a low carbon content are examples of materials having high permeability.

Materials of low permeability retain a strong residual magnetism after the magnetizing force is removed. These materials are hard to magnetize. Very hard ferromagnetic steel, like that of a horseshoe magnet, has high retentivity . . . it retains a strong residual magnetic field.

Turn ahead to page 1-64.

From page 1-64

**Coercive force or coercivity** is defined as:

"THE REVERSE MAGNETIZING FORCE REQUIRED TO REDUCE THE RESIDUAL MAGNETISM TO ZERO."

Which of the following would require the strongest coercive force to remove residual magnetism?

Iron . . . . . . . . . . . . . . . . . . . . . . . . . . . . . . . . . . . . . . . . . Page 1-65
Hardened ferromagnetic steel . . . . . . . . . . . . . . . . . . . . . Page 1-67

Absolutely. Material of low permeability (hard ferromagnetic steel) would have the greatest retentivity. It would retain the strongest residual magnetism . . . like a permanent magnet.

If the magnetizing force is now reversed, as is the case with alternating current, and gradually increased in the reversed direction, the residual flux density is reduced to zero at point c by a magnetizing force represented by the distance 0 - c.

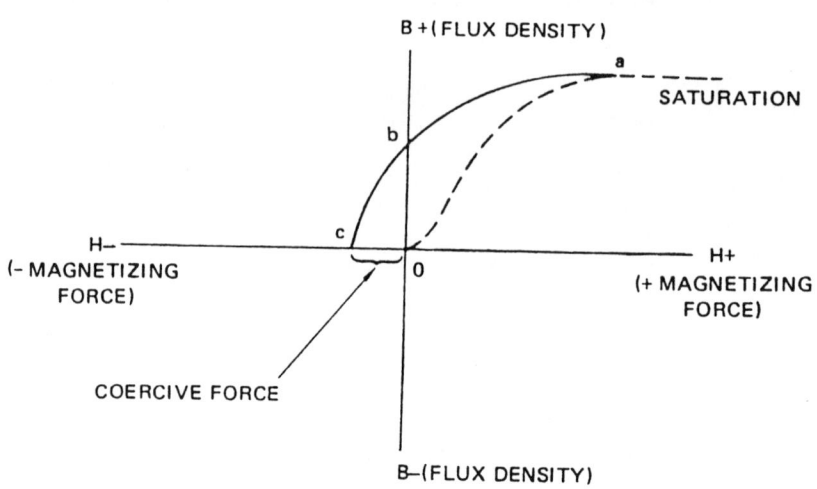

With flux density reduced to zero at point c, the material is demagnetized and we can determine the *coercive force* for the material.

Turn back to page 1-63.

You think iron would require the strongest coercive force to remove the residual magnetism. Don't forget now, iron is soft in comparison to steel. Iron, particularly very soft iron, retains or holds only a small amount of residual magnetism after the magnetizing force is removed. Here is the definition again.

**Coercive force** is defined as:

"THE REVERSE MAGNETIZING FORCE REQUIRED TO REDUCE THE RESIDUAL MAGNETISM TO ZERO."

Since hard steel has high retentivity—retains a strong residual magnetic field—don't you agree that it would require the strongest coercive force to remove the residual magnetism?

Turn ahead to page 1-67.

From page 1-63                                                    1-67

Yep, that's right.  Hardened ferromagnetic steel would require the strongest coercive force.  In other words, the steel would require the strongest reverse magnetizing force to remove the residual magnetism. The steel also has the greatest retentivity—retains the strongest residual magnetic field.

As the reverse magnetizing force is increased beyond point c, flux density increases to the saturation point *in the reverse direction*—point d.

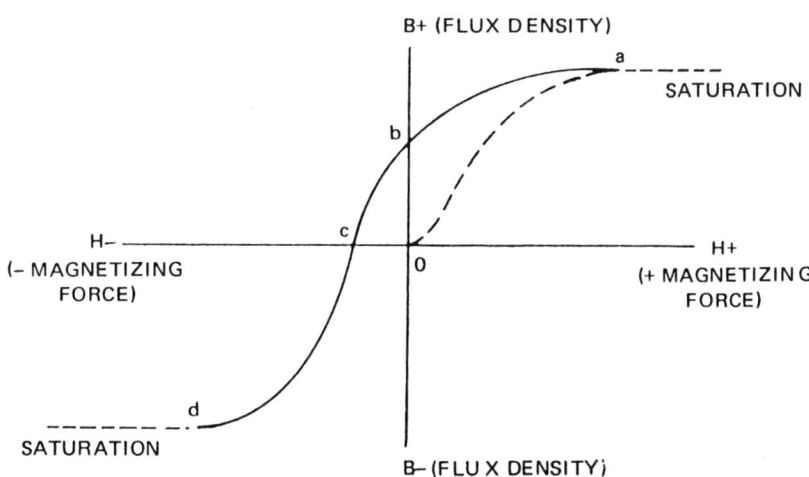

We have defined **coercive force** as "the reverse magnetizing force required to reduce the residual magnetism to zero."

**Between which of the following points on the curve is the coercive force shown?**

Between points 0 (zero) and b .................... Page 1-68
Between points 0 (zero) and c .................... Page 1-70

From page 1-70                                                                 1-66

You selected "residual magnetism." It isn't the answer we were after, but you are right although a little incomplete. If we reduce the magnetizing force to zero (point e) in the reversed direction there will be residual magnetism . . . except that it will be reverse residual magnetism.

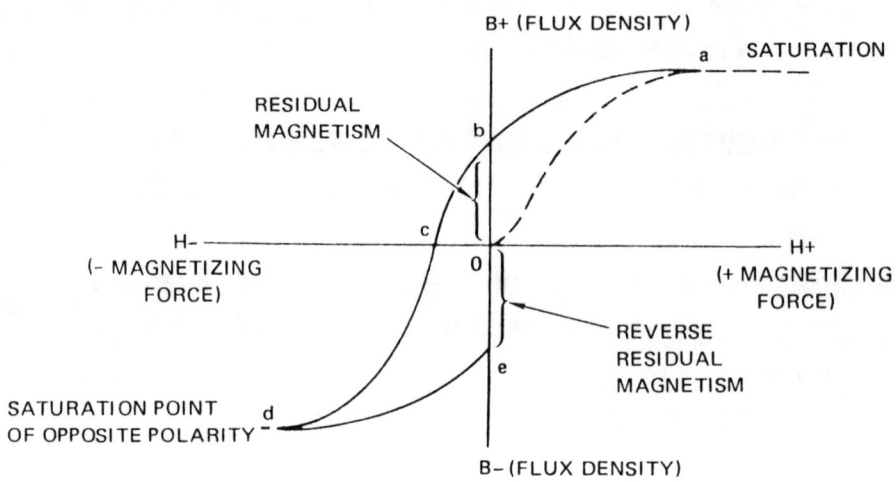

Here you can see the two areas representing residual magnetism. The re-reduction of the magnetizing force to zero at point e results in reverse magnetism.

Turn ahead to page 1-69.

You think the coercive force is shown between points 0 (zero) and b. Let's take another look at the curve and see if you still feel that way.

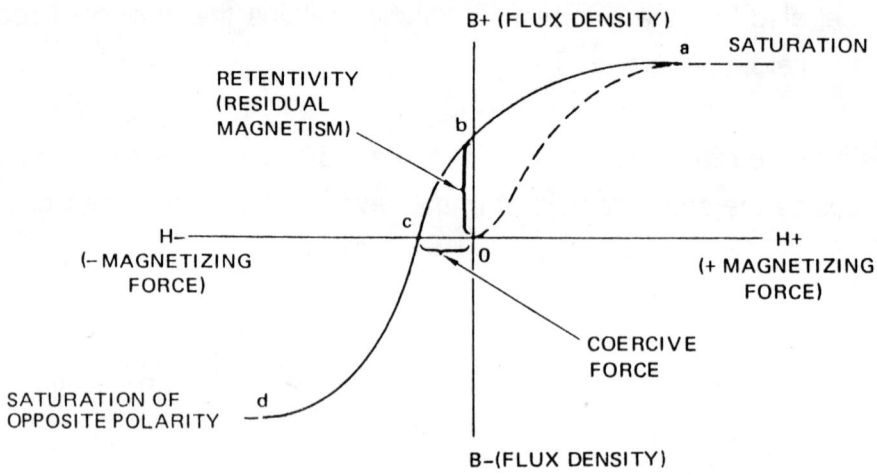

Along the dashed line, flux density increases until the material is saturated (point a). When the magnetizing force is reduced to zero (point b) we can measure the residual magnetism as shown. If the magnetizing force is now reversed and gradually increased in the reversed direction, flux density is reduced to zero at point c. It is between points c and 0 that we measure the coercive force required to eliminate or remove the residual magnetism from the material.

Remember, **coercive force** is defined as:

**"THE REVERSE MAGNETIZING FORCE REQUIRED TO REDUCE THE RESIDUAL MAGNETISM TO ZERO."**

So you see, the coercive force is shown between point 0 (zero) and c.

Turn ahead to page 1-70.

From page 1-70

Of course. If we again reduce the magnetizing force to zero (point e) in the reverse direction, we would have reverse residual magnetism. Point e also shows the retentivity or remanence—the ability of the material to retain residual magnetism.

By increasing the magnetizing force in the original direction we complete a loop. Notice, however, that the dashed line is no longer followed. It was the "virgin" curve or first curve.

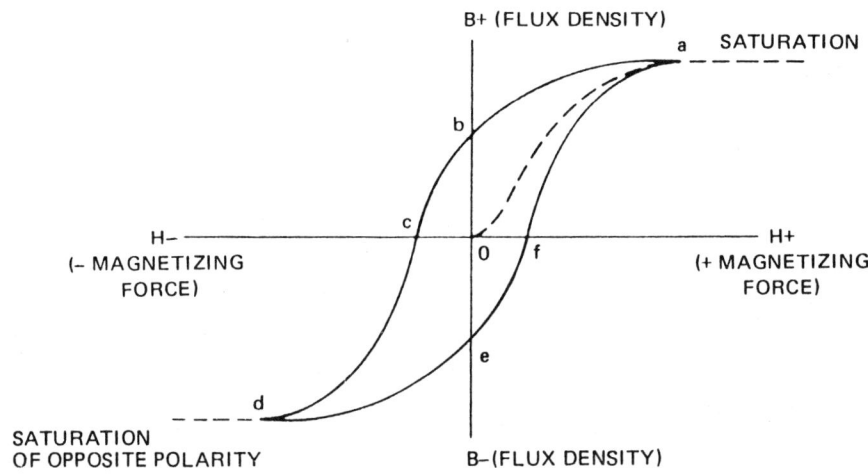

Having established a residual magnetic field in the reverse direction, it will be necessary to remove it. The force required to remove this reverse residual field is shown between points 0 and f.

**What is the name of the reverse magnetizing force required to remove residual magnetism from the material?**

**Retentivity** .................................. Page 1-71
**Coercivity** .................................. Page 1-72
**Magnetizing force** .......................... Page 1-74

From page 1-67

Good for you.  Coercive force is shown between points 0 (zero) and c. The other area, points 0 and b, represented the residual magnetism or *retentivity* of the material.

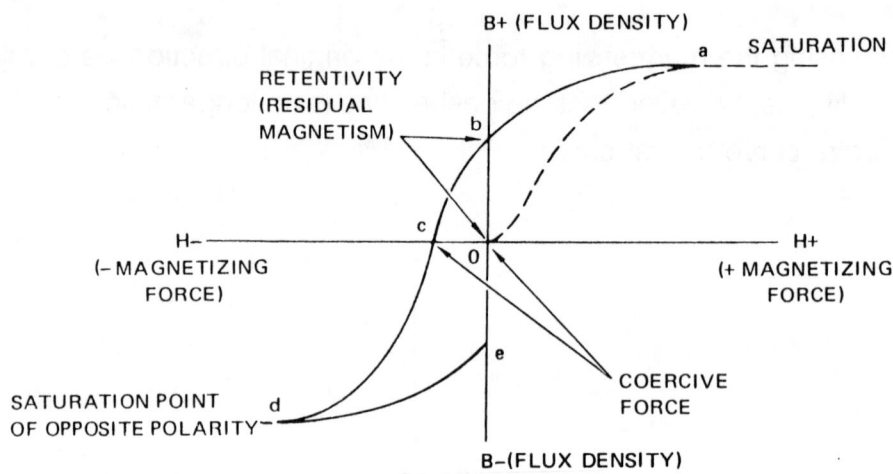

Moving left on the H axis, point d on the curve is the point of maximum saturation in the reverse direction. In other words, the material has been magnetized to its maximum flux density in the reverse direction, down on the B axis.

**If we again reduce the magnetizing force to zero (point e), which of the following do you think will exist in the material?**

**Residual magnetism** .............................. Page 1-66
**Reverse residual magnetism** ..................... Page 1-69

You selected "retentivity." No, that isn't the name of the reverse magnetizing force required to remove the residual magnetism from the material.

**Retentivity** is defined as:

> **"THE ABILITY OF A MATERIAL TO RETAIN A PORTION OF THE MAGNETIC FIELD SET UP IN IT AFTER THE MAGNETIZING FORCE HAS BEEN REMOVED."**

Return to page 1-69 and try again.

From page 1-69

Yes, coercivity is the correct answer. That is the name of the reverse magnetizing force required to remove the residual magnetism from the material. In this case, the coercive force is shown between points 0 and f. Note that the coercive force 0 - f is equal *but in opposite direction* to coercive force 0 - c.

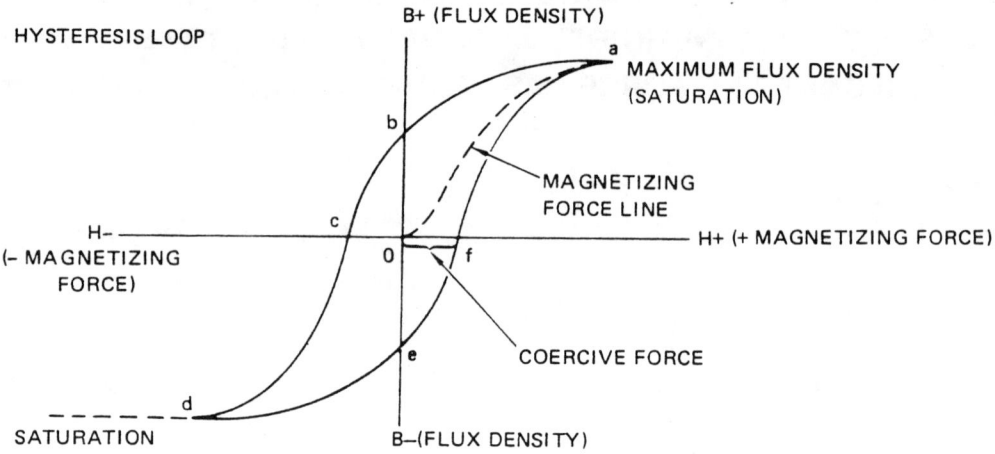

The closed loop that has been formed is called a "hysteresis loop." The hysteresis loop gets its name from the lag between the magnetizing force and the increase of flux density throughout the cycle. This lag is called *hysteresis*. The lag is shown between points 0 and f.

Turn to the next page.

From page 1-72

A material that is hard to magnetize is said to have *high reluctance*.

**RELUCTANCE** is defined as:

> "THE RESISTANCE OF A MATERIAL TO CHANGES IN MAGNETIC FIELD STRENGTH."

A hardened piece of ferromagnetic steel is hard to magnetize but would retain a strong residual magnetic field. If a hysteresis loop was plotted for this steel, what would happen to the distance between points 0 and f?

**In other words, would the coercive force be stronger or weaker?**

**Stronger** .................................. **Page 1-75**
**Weaker** ................................... **Page 1-76**

From page 1-69

You think "magnetizing force" is the name of the reverse magnetizing force required to remove the residual magnetism from the material. No, what we are after is the name of a specific portion of the magnetizing force. Here is the curve again.

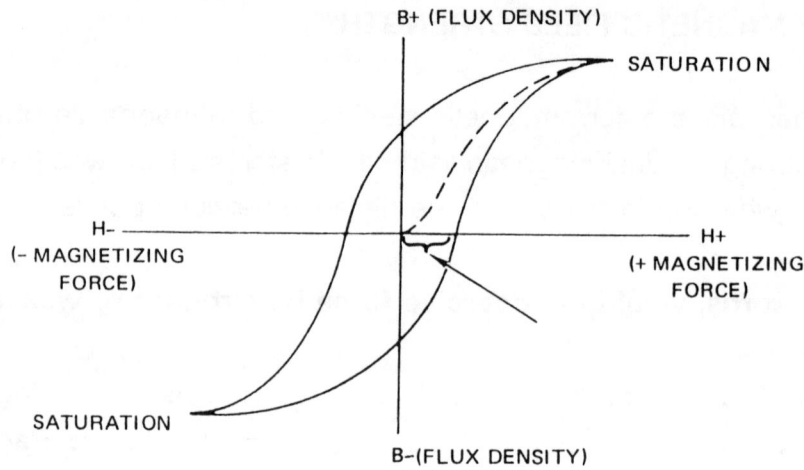

The arrow is pointing at the specific portion of the magnetizing force we are talking about. This area shows the REVERSE MAGNETIZING FORCE REQUIRED TO REMOVE RESIDUAL MAGNETISM FROM THE MATERIAL.

Turn back to page 1-69 and select the correct name for this area.

From page 1-73

Stronger is the right answer. Because a hardened piece of ferromagnetic steel would retain a strong residual magnetic field, the reverse magnetizing force required to remove the residual magnetism would have to be stronger. Here is a typical hysteresis loop for a hardened piece of ferromagnetic steel.

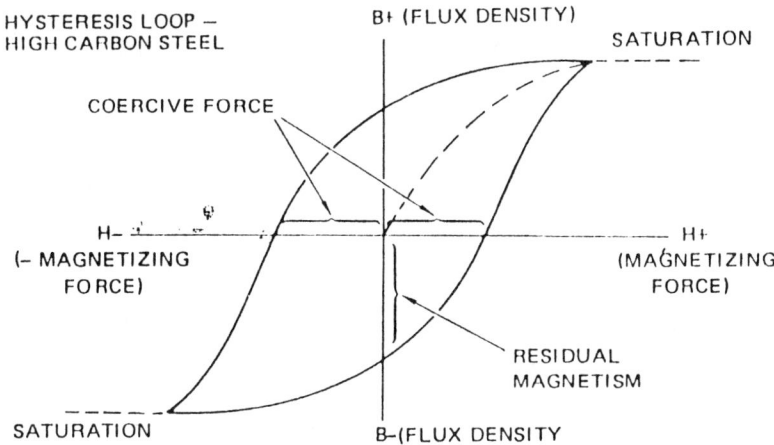

Here you can see that the coercive force would have to be stronger because the residual magnetic field in the part would be stronger. A wide hysteresis loop indicates a material that is difficult to magnetize—has high reluctance. Reluctance is the inverse of permeability. Permeability also can be defined as the ratio of flux density to magnetizing force on the hysteresis curve.

Turn ahead to page 1-77.

From page 1-73

You're guessing. You think that the coercive force would be weaker for a hardened piece of steel? Let's review the characteristics of hardened ferromagnetic steel.

- Hardened steel has LOW PERMEABILITY . . . it is hard to magnetize.

- Hardened steel has HIGH RETENTIVITY . . . it retains a strong residual field.

- Hardened steel has HIGH RELUCTANCE . . . it has high resistance to changes in magnetic field strength.

In other words, with high reluctance, hardened ferromagnetic steel would require a stronger coercive force.

Turn back to page 1-75.

In short, a wide hysteresis loop shows that hardened steel would have the following qualities:

- LOW PERMEABILITY . . . hard to magnetize.

- HIGH COERCIVE FORCE . . . requires a high reverse magnetizing force to remove residual magnetism.

- HIGH RELUCTANCE . . . high resistance to changes in magnetic field strength.

- HIGH RESIDUAL MAGNETISM . . . retains a strong residual magnetic field.

Turn to the next page.

A thin hysteresis loop indicates a material of low retentivity.

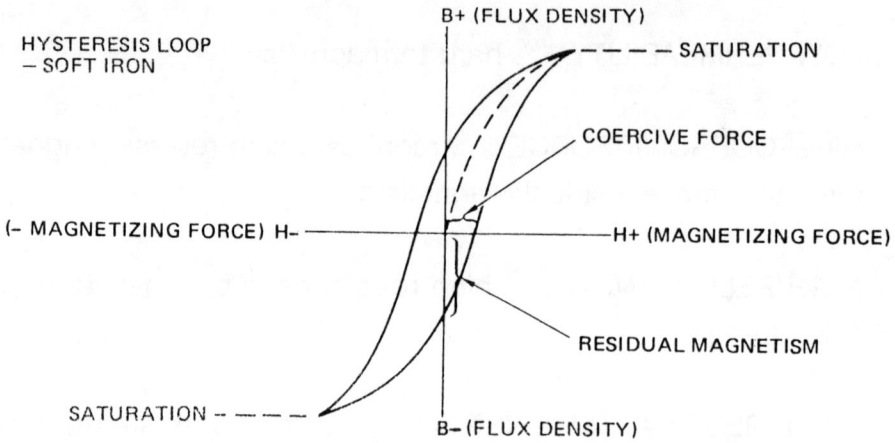

This hysteresis loop shows the qualities of a soft material like iron with a low carbon content. The coercive force is low because the material retains only a weak residual magnetic field. In short, this loop shows that soft iron would have the following qualities:

- HIGH PERMEABILITY . . . easy to magnetize.

- LOW COERCIVE FORCE . . . requires a low reverse magnetizing force to remove residual magnetism.

- LOW RELUCTANCE . . . low resistance to changes in magnetic field strength.

- LOW RESIDUAL MAGNETISM . . . retains a weak residual magnetic field.

Turn to the next page.

From page 1-78

Perhaps you have already recognized that the hysteresis loop shows what occurs during one cycle of alternating current. If the current value is not changed, the magnetic field follows the outside perimeter of this curve as long as the alternating current is flowing.

Take a few minutes to review any new terms and the hysteresis (or magnetization) curve before turning to page 1-80 for a review of what we've covered so far.

## CHAPTER REVIEW

_A_  1.  By definition, ferrous means "of or pertaining to _____."

   A. iron
   B. a type of wheel
   C. magnetic particle examination
   D. magnetism

_C_  2.  A wide hysteresis loop indicates a material that is difficult to magnetize. The material has high:

   A. permeability.
   B. flux density.
   C. reluctance.
   D. saturation.

_A_  3.  A piece of copper wire will not be attracted to a magnet and it is called a _____ material.

   A. nonmagnetic
   B. magnetic
   C. north pole
   D. domain

From page 1-80

_B_  4.  All ferromagnetic materials are attracted to a:

   A.  domain.
   B.  magnet (magnetic field).
   C.  nonferromagnetic material.
   D.  permeable flux.

_C_  5.  When a magnetizing force (current) is removed, the material may retain some of the:

   A.  electricity.
   B.  leakage.
   C.  magnetic field (magnetism).
   D.  "hardness."

_A_  6.  Some materials are harder to magnetize than others. This is caused by the _____ of the material.

   A.  permeability
   B.  color
   C.  density
   D.  length

A    7.    _____ is defined as the ease with which materials can be magnetized.

   A.   Permeability
   B.   Reluctance
   C.   Remanence
   D.   Ferromagnetic

D    8.    Point "a" on the hysteresis loop shows the maximum flux density for the steel. In other words, the steel is:

   A.   magnetic.
   B.   useless.
   C.   hardened.
   D.   saturated.

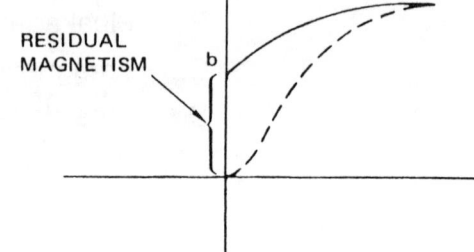

B    9.    Hardened ferromagnetic steel is more difficult to magnetize, so it is said to possess _____ permeability.

   A.   high
   B.   low
   C.   average
   D.   dominant

From page 1-82                                                          1-83

_B_     10.   A nail is made of iron so it is made of _____ material.

              A.   round
              B.   ferrous
              C.   nonferrous
              D.   soft

_B_     11.   _____ magnetism is defined as the magnetic field which remains in a material after the magnetizing force is removed.

              A.   Permeable
              B.   Residual
              C.   Nonferromagnetic
              D.   Horseshoe

_D_     12.   Soft iron is easy to magnetize but does not retain very much magnetism. It has _____ permeability but retains a _____ residual magnetic field.

              A.   low, strong
              B.   low, weak
              C.   high, strong
              D.   high, weak

From page 1-83

___C___ 13. Hardened ferromagnetic steel is more difficult to magnetize, but when the current is shut off, the steel retains most of its magnetism. Hardened ferromagnetic steel has _____ permeability but retains a _____ residual magnetic field.

    A.    high, weak
    B.    high, strong
    C.    low, strong
    D.    low, weak

___B___ 14. The ease with which a ferromagnetic material can be magnetized is a measure of its:

    A.    elasticity.
    B.    permeability.
    C.    flux density.
    D.    magnetite level.

___A___ 15. Soft iron is very easy to magnetize and has high:

    A.    permeability.
    B.    remanence.
    C.    flux capacity.
    D.    density.

From page 1-84                                                          1-85

_B_    16.   The magnetic field that remains in the metal after the magnetizing force is removed is called _____ magnetism.

             A.   extra
             B.   residual
             C.   slight
             D.   permeable

_A_    17.   Residual magnetism is always less than the magnetic field which is present when the magnetizing _____ is on.

             A.   force (current)
             B.   magnetite
             C.   diamagnetic
             D.   alnico

_B_    18.   A ferromagnetic material that has low permeability will retain a _____ magnetic field.

             A.   weak
             B.   strong

_A_    19.   A horseshoe magnet that is made of hardened ferromagnetic material will retain a _____ residual field.

             A.   strong
             B.   weak
             C.   long
             D.   wide

From page 1-85

B  20. The horseshoe magnet that is made of hardened ferromagnetic material also has _____ permeability.

   A. high
   B. low
   C. long
   D. wide

B  21. The permeability of a specific material can be determined by its _____ loop.

   A. magnetic
   B. hysteresis
   C. flexibility
   D. coercivity

A  22. The ratio of flux density (B) to magnetizing force (H) equals the _____ of the material.

   A. permeability
   B. flexibility
   C. coercivity
   D. magnetite

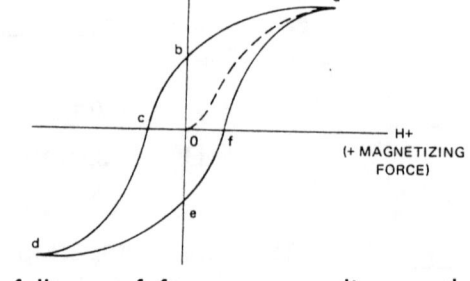

C  23. The total number of lines of force per unit area is called:

   A. coercive force.
   B. magnetic distance.
   C. flux density.
   D. linear area.

B  24. Magnetic saturation for the material is shown at point "a." In other words, point "a" shows the maximum number of lines of force the material will hold. It is the point of maximum:

A. permeability.
B. flux density.
C. magnetizing force.
D. hysteresis.

A  25. If the magnetizing force is reduced to zero (point b), we can measure the residual magnetism or _____ of the material.

A. retentivity
B. flux density
C. coercivity
D. hardness

B  26. Hardened ferromagnetic steel retains a strong residual magnetic field, so it has high:

A. permeability.
B. retentivity.
C. saturation.
D. flux density.

_C_ 27. If the magnetizing force is reversed and gradually increased in the reversed direction, flux density is reduced to zero at point c. The distance between points c and 0 (zero) measures the:

A. flux density.
B. remanence.
C. coercive force.
D. permeability.

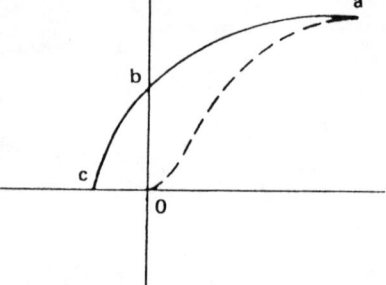

_A_ 28. Coercive force is defined as the reverse magnetizing force required to remove any _____ from the material.

A. residual magnetism
B. permeability
C. ferromagnetite
D. magnetizing current

_B_ 29. As the reverse magnetizing force is increased, flux density increases to the saturation point in the reverse direction (point d). This is a point of maximum:

A. residual magnetism.
B. flux density.
C. permeability.
D. magnetizing force.

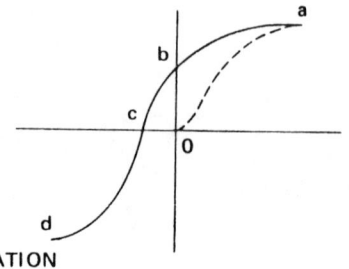

SATURATION POINT

A  30. By increasing the magnetizing force in the original direction from point e to point f, we can measure the reverse magnetizing force required to remove the _____ from the material.

   A. residual magnetism
   B. permeability
   C. magnetizing force
   D. magnetizing current

C  31. The distance between points 0 (zero) and f represents the magnetizing force required to remove the residual magnetism from the material. This force is called the _____ force.

   A. saturation
   B. magnetizing
   C. coercive
   D. ferromagnetic

B  32. Reluctance is defined as the resistance of a material to changes in _____ field strength.

   A. electrical
   B. magnetic
   C. optical
   D. coercive

From page 1-89

__C__   33.  If we again reduce the magnetizing force to zero (point e) in the reverse direction, we can measure the reverse:

  A.  magnetizing force.
  B.  permeability.
  C.  residual magnetism.
  D.  magnetizing current. SATURATION POINT

__B__   34.  A hardened piece of ferromagnetic steel is hard to magnetize and is said to have high reluctance and retains a strong residual magnetic field. Removal of the residual field would require a(n) _____ coercive force.

  A.  weak
  B.  strong
  C.  minimal
  D.  average

__D__   35.  A thin hysteresis loop indicates a material that is easy to magnetize. The material has _____ reluctance.

  A.  high
  B.  average
  C.  strong
  D.  low

36. A nail is made of iron which is ferrous metal. Since the nail is attracted to a magnet, it is called _____ material.

   A. permeable
   B. nonferromagnetic
   C. a polar
   D. ferromagnetic

37. The ability of material to retain a certain amount of residual magnetism is the definition of:

   A. permeability.
   B. retentivity.
   C. coercivity.
   D. flux density.

38. Electric current is used to create a magnetic field in a _____ material.

   A. nonferromagnetic
   B. saturated
   C. ferromagnetic
   D. hysteresis prone

C   39. The magnetic field which remains in a material after the magnetizing force is removed is the definition of:

   A. permeability.
   B. coercivity.
   C. residual magnetism.
   D. flux density.

D   40. The reverse magnetizing force required to remove residual magnetism from the material is the definition of:

   A. remanence.
   B. permeability.
   C. flux density.
   D. coercive force.

C   41. The resistance of a material to changes in magnetic field strength is the definition of:

   A. permeability.
   B. coercivity.
   C. reluctance.
   D. flux density.

B    42.  The number of lines of force per unit area is the definition of:

A. permeability.
B. flux density.
C. reluctance.
D. remanence.

## ANSWERS TO REVIEW QUESTIONS
## FOR CHAPTER 1

| Question & Answer | | Reference Page(s) |
|---|---|---|
| 1. | A | 1-38 |
| 2. | C | 1-77 |
| 3. | A | 1-44 |
| 4. | B | 1-44 |
| 5. | C | 1-43 |
| | | |
| 6. | A | 1-47 |
| 7. | A | 1-46 |
| 8. | D | 1-59 |
| 9. | B | 1-43 |
| 10. | B | 1-42 |
| | | |
| 11. | B | 1-43 |
| 12. | D | 1-78 |
| 13. | C | 1-77 |
| 14. | B | 1-46 |
| 15. | A | 1-47 |
| | | |
| 16. | B | 1-48 |
| 17. | A | 1-53 |
| 18. | B | 1-53 |
| 19. | A | 1-53 |
| 20. | B | 1-43 |

| Question & Answer | Reference Page(s) |
|---|---|
| 21. B | 1-77, 78 |
| 22. A | 1-75 |
| 23. C | 1-51 |
| 24. B | 1-59 |
| 25. A | 1-62 |
| | |
| 26. B | 1-62 |
| 27. C | 1-68 |
| 28. A | 1-63 |
| 29. B | 1-67 |
| 30. A | 1-69 |
| | |
| 31. C | 1-72 |
| 32. B | 1-73 |
| 33. C | 1-69 |
| 34. B | 1-77 |
| 35. D | 1-78 |
| | |
| 36. D | 1-38 |
| 37. B | 1-61 |
| 38. C | 1-53 |
| 39. C | 1-43 |
| 40. D | 1-63 |
| | |
| 41. C | 1-73 |
| 42. B | 1-49 |

# CHAPTER 2

## PRODUCING MAGNETIC FIELDS

As we have seen, there is much left to consider concerning magnetism and its theory. We have discussed only the most basic concepts. It is most important for us to note how magnetic fields behave within and near the surface of a ferromagnetic test article in the area of a discontinuity.

Permanent magnets were found to be useful sources of magnetic fields, but it will usually take a larger field to provide the strength needed for most magnetic particle examinations. Magnetic fields occur anytime an electrical current passes through an electrical conductor. We began to observe this in our residual magnetism discussion using a typical battery. Let us explore ways we can utilize electrical currents and the resulting powerful magnetic field in magnetic particle testing.

### Magnetizing Current

With electric current flowing through a copper wire, a magnetic field is created around the wire. The magnetic lines of force are always at right angles to the direction of electric current flow. In this case, the lines of force are traveling counterclockwise around the wire.

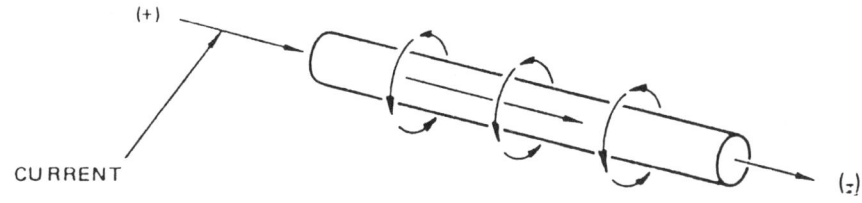

Turn to the next page.

It is important to note that this text will consider that electricity flows from positive (+) to negative (-). This is typically referred to as "current" flow.

You may study other texts (i.e. welding) that will refer to electricity flowing from negative (-) to positive (+). This is referred to as "electron" flow. However, don't let these different definitions confuse you. The basic principles remain the same regardless of how you define the direction of flow.

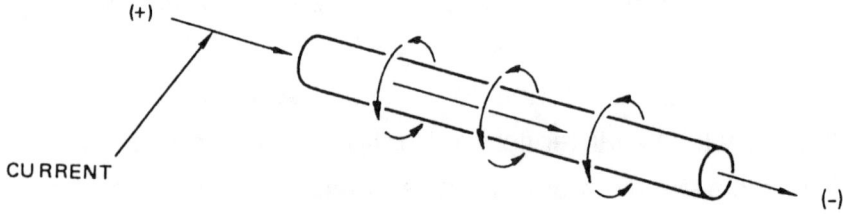

Here is a simple rule to assist you in determining the direction of the lines of force. Imagine that you grasp the wire with your right hand so that your thumb points in the direction of electric "current" flow. Your fingers will now be pointing in the direction of flow of the lines of force. This method of determining the direction of lines of force is known as the *right-hand rule*.

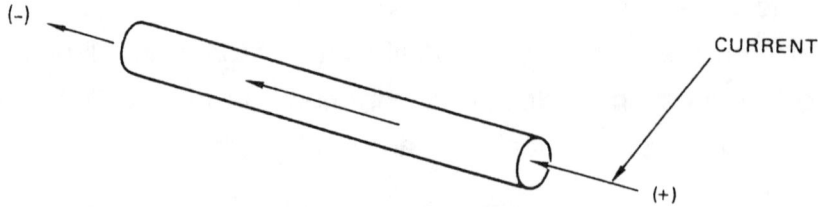

**Using the right-hand rule, what is the direction of flow of the lines of force around this wire?**

**Counterclockwise** .............................. **Page 2-4**
**Clockwise** .................................. **Page 2-7**

From page 2-7

Right. Using the right-hand rule, the lines of force would be turning *clockwise* around the wire. The most important point in this discussion however is that the magnetic field created around the wire will be at *right angles* or *perpendicular* to the wire.

Again using the right-hand rule, if we reverse the direction of current flow, we would also reverse the direction of the circular field; but the magnetic field generated by the current through the wire would remain perpendicular to the wire.

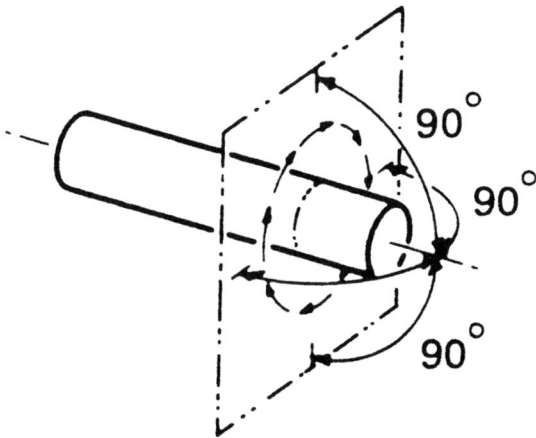

At every point along the wire there would be a circular magnetic field that is perpendicular to the wire.

**Therefore, we don't really care what direction the current is flowing through the wire.**

**True** .................................... **Page 2-6**
**False** ................................... **Page 2-8**

From page 2-2

Counterclockwise is not correct. Perhaps we've confused you as to how we are looking at the wire. Let's look at it this way.

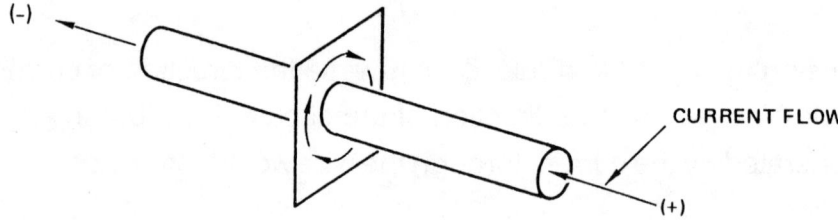

The wire has been passed through a piece of paper to give you a better idea of the direction of flow of the lines of force. They are flowing clockwise 90° to the direction of electric current flow. Return to page 2-2 and study the drawing again.

Counterclockwise isn't the right answer. This diagram may help.

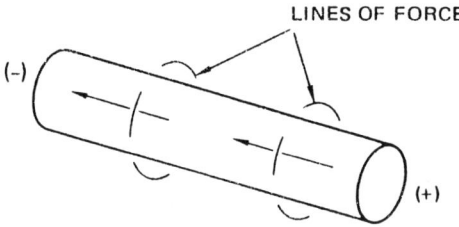

Here the electric current is flowing into the right end of the wire at the positive (+) end and out of the wire on the left which is the negative (-) end. The question is, which way are the magnetic lines of force turning?

Apply the right-hand rule and turn back to page 2-3.

From page 2-3                                                                 2-6

Right! We do not care what direction the current is flowing so long as we know that the magnetic field generated is perpendicular to the current (or wire). Just remember that the field established by the current is always *circular and at 90°* to the current. This leads us to our next topic.

## Circular Magnetic Fields

A CIRCULAR FIELD is established *in and around* any electrically-conductive material when a current is passed through it.

Suppose we substitute an iron bar in place of the copper wire and pass a current through it. The magnetic field will look something like the illustration below.

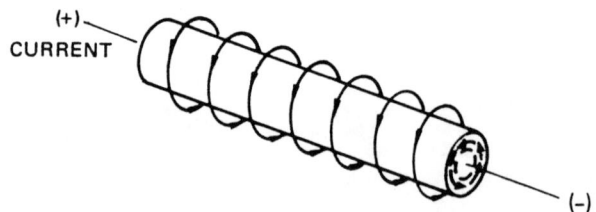

**From this illustration where is the magnetic field located?**

Only inside the bar . . . . . . . . . . . . . . . . . . . . . . . . . . . . . . . . . **Page 2-9**
Only around the bar . . . . . . . . . . . . . . . . . . . . . . . . . . . . . . . **Page 2-10**
In AND around the bar . . . . . . . . . . . . . . . . . . . . . . . . . . . . . **Page 2-13**

From page 2-2

That's absolutely correct. The lines of force would be traveling around the wire clockwise like this.

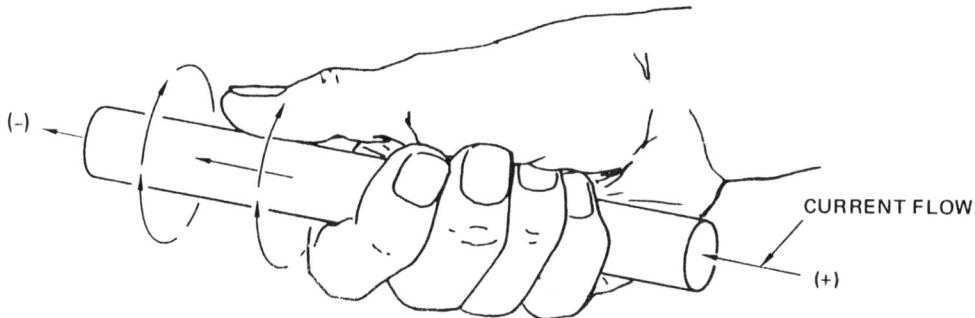

By grasping the wire in the right hand with the thumb pointing in the direction of current flow, the fingers will point in the same direction as the lines of force. Now, let's take a look at an end view of the wire.

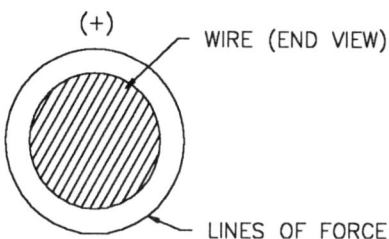

In this view, we can see the magnetic lines of force around the wire. The plus (+) sign means that we are looking at the positive end of the wire.

**Using the right-hand rule, what would be the direction of flow of the lines of force around the wire?**

**Clockwise** . . . . . . . . . . . . . . . . . . . . . . . . . . . . . . . . . . . . . . . **Page 2-3**
**Counterclockwise** . . . . . . . . . . . . . . . . . . . . . . . . . . . . . . . . **Page 2-5**

From page 2-3

We have confused you. We really *do not* care which direction the current is flowing.

Do not confuse the direction of the lines of force with the location of the field. The lines of force may be flowing either clockwise or counterclockwise around the wire, but the field in either case is at right angles to the current flow.

We will, in fact, state this as a rule . . . **IN CIRCULAR MAGNETIZATION THE FIELD PRODUCED IS PERPENDICULAR TO THE CURRENT FLOW**.

Turn back to page 2-6.

From page 2-6

You felt that the circular magnetic field would be located *only inside* the bar. Let's look again at the illustration of the field formed by passing a current through the iron bar.

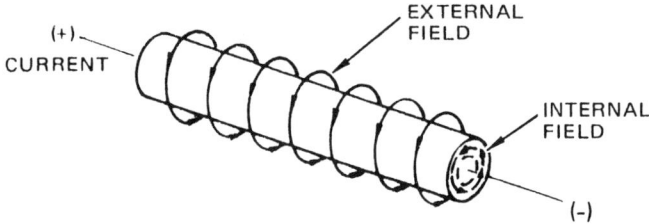

The magnetic field that is established by the current lies within the bar AND *around* the bar.

Remember—a CIRCULAR FIELD is established *in and around* any electrically-conductive material when a current is passed through it. This applies to copper rods and also iron bars.

Turn ahead to page 2-13.

You felt that the circular magnetic field would be located *only* around the bar. Let's look at it again.

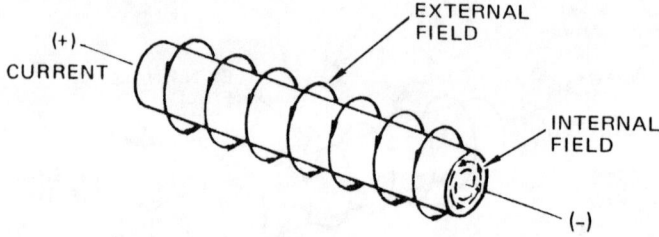

The magnetic field that is established by the current lies *within AND around* the bar.

Remember—a circular field is established *in AND around* any electrically-conductive material when a current is passed through it. This applies to copper rods as well as iron bars.

Turn ahead to page 2-13.

From page 2-13

2-11

You must remember: 1) iron particles will only be attracted to magnetic poles, and 2) magnetic poles are formed only where magnetic flux *leaves or enters* the material. Here's the circular magnet once again.

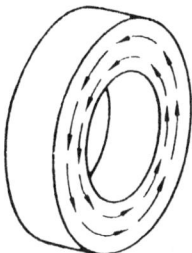

The magnetic field is contained entirely within the magnet. No lines of force enter or leave it; therefore, no magnetic poles are formed and iron filings will not be attracted.

If the field is not disturbed by any cracks or other discontinuities, the circular field is leakage free—no poles are formed.

Turn ahead to page 2-14.

From page 2-14

Exactly so. The crack will cause flux leakage in the circularly-magnetized bar just as it did in the circular magnet.

Now, let's magnetize this rectangular bar. The *internal* field is still called a "circular" field even though the shape of the bar causes the field to look like this.

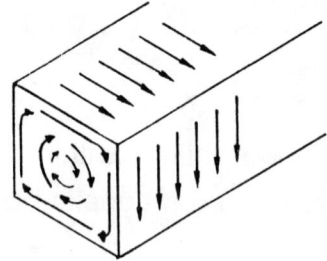

Here we show a crack in the rectangularly-shaped bar.

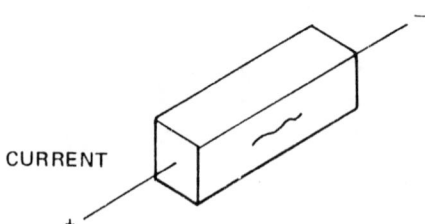
CURRENT

**Do you think this crack will cause a flux leakage by interrupting the *internal* circular field?**

**Yes** .................................................. **Page 2-15**
**No** ................................................... **Page 2-17**

Excellent! The circular magnetic field is established *in and around* the ferromagnetic iron bar since it is an electrically-conductive material. The magnetic field around a copper rod carrying current will also be in and around the rod; however, there is a distinct difference between the fields established.

Since copper is nonmagnetic, the field distribution is not the same as the field distribution for the iron bar which is magnetic. The magnetic material is more strongly magnetized and holds more of the field within the bar. Therefore, the field strength *within* the magnetic bar is greater than the field strength *within* the copper rod. The fields outside the two are practically identical. We'll go into more detail about that a little later in the program.

For now, let's examine CIRCULAR FIELDS.

Remember the circular magnet?

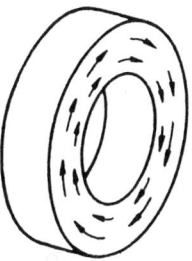

**If we dust iron particles on this circular magnet would they be attracted to any point on the magnet?**

**Yes** .................................................. **Page 2-11**
**No** ................................................... **Page 2-14**

From page 2-13                                                            2-14

You are absolutely right. Iron particles would *not* be attracted to the circular magnet. Iron particles would only be attracted where there are magnetic poles caused by magnetic flux leakage either leaving or entering the magnet. If the circular magnet had a crack in it, iron particles would be attracted at the flux leakage.

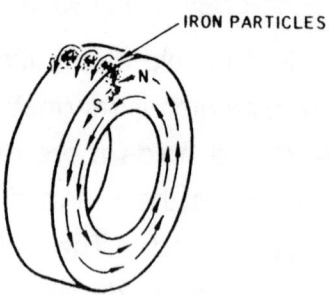

Now, if we were to pass current through a cylindrically-shaped iron bar, we would obtain a circular magnetic field *in and around* the bar as illustrated below in View A.

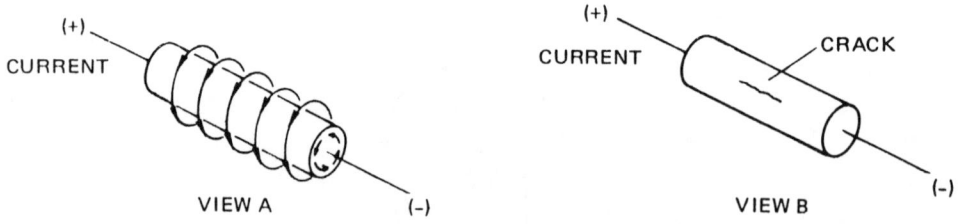

This field, like a circular magnet, is leakage free. In View B we show an identical bar with a crack in it.

**When a circular field is established in the bar by passing current through it, can we expect the crack to cause a flux leakage by interrupting the circular field?**

**Yes** .................................................... Page 2-12
**No** ..................................................... Page 2-16

From page 2-12    2-15

Right! The crack in the rectangularly-shaped bar will cause flux leakage. North and south poles will be formed when the crack forces some of the lines of force to leave the bar. These magnetic poles will attract iron filings.

And that is exactly how cracks are located with magnetic particle inspection. If our square steel bar has a crack in the surface 90° to the direction of the lines of force within the bar, iron particles will be attracted to the crack.

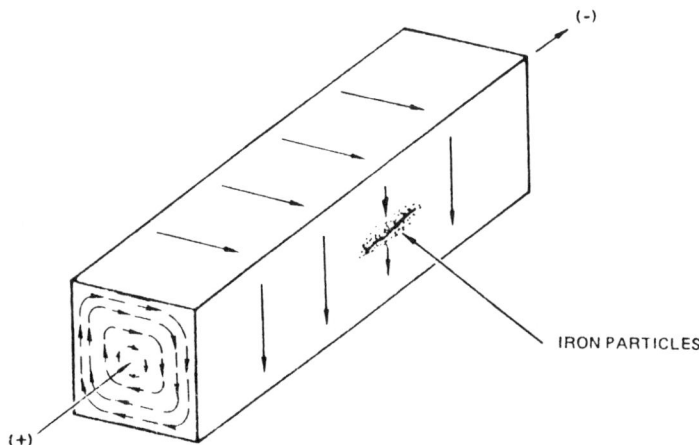

The crack in the bar has caused a north and south pole. Some of the lines of force have been forced to the surface creating *flux leakage*. The flux leakage attracts the iron particles.

Turn ahead to page 2-18.

Perhaps you are confused by the field around the bar. It is true that the crack will have no effect on the field surrounding the bar but the crack will affect the field *within* the bar.

We can ignore the external field since it has no magnetic poles. Let's look at what happens to the internal field in this enlarged view.

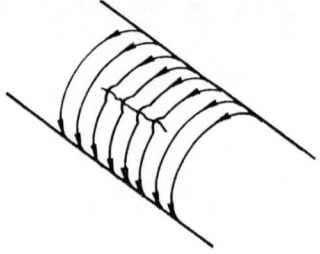

The crack forces the lines of flux to leave the material and re-enter it. As you know, this causes magnetic poles to be formed. In short, the crack will cause flux leakage from the field *inside* the iron bar.

Turn back to page 2-12.

From page 2-12

We've confused you in some way. The crack *will* cause flux leakage since it interrupts the circular field *within* the bar.

Here we show what is happening at the crack.

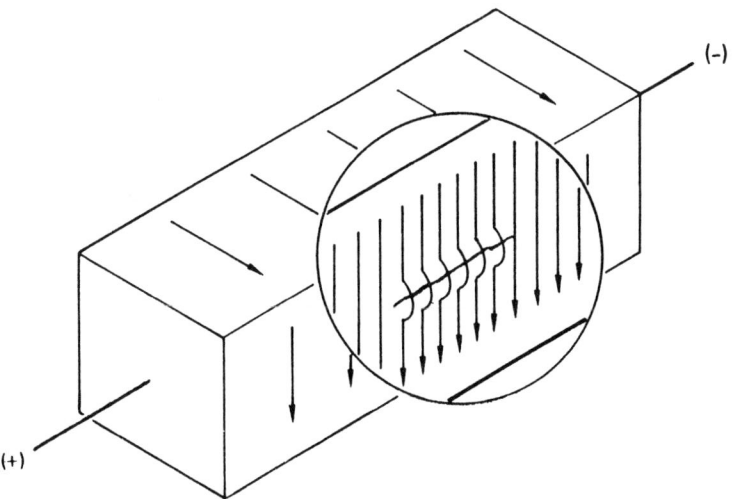

The circular magnetic field is interrupted by the crack even though the bar is rectangular in shape.

Turn back to page 2-15.

From page 2-15

*Circular magnetization* will detect cracks that are *between* 45° and 90° to the lines of force as illustrated below.

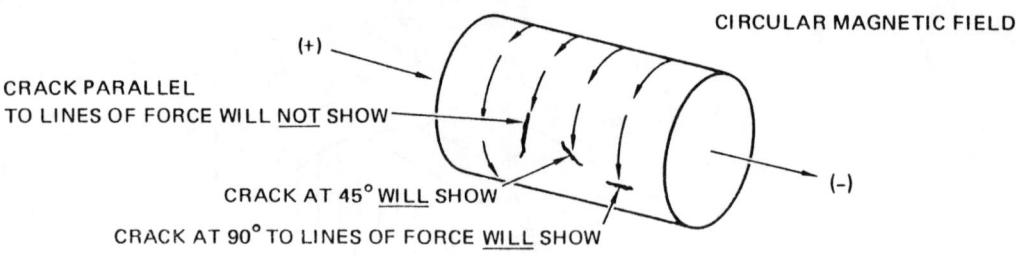

The crack that runs crosswise or 90° to the lines of force will have the most flux leakage and will attract the most iron particles. The crack at 45° will also have flux leakage and will attract iron particles.

The crack that runs parallel to the lines of force does not present enough of the crack area to disrupt the lines of force and will generally not cause enough flux leakage to attract iron particles. The lines of force tend to remain in the metal and squeeze around this crack.

**Any crack that runs parallel to the lines of force will NOT attract iron particles. Which of the following explains why this is so?**

**The lines of force pass through the crack** . . . . . . . . . . . . . **Page 2-20**
**No poles or flux leakage exist at the crack** . . . . . . . . . . . **Page 2-23**
**The crack is at 90° to the lines of force** . . . . . . . . . . . . . . **Page 2-24**

From page 2-23                                                                                                        2-19

Excellent. That's right. Crack A will NOT attract iron particles.

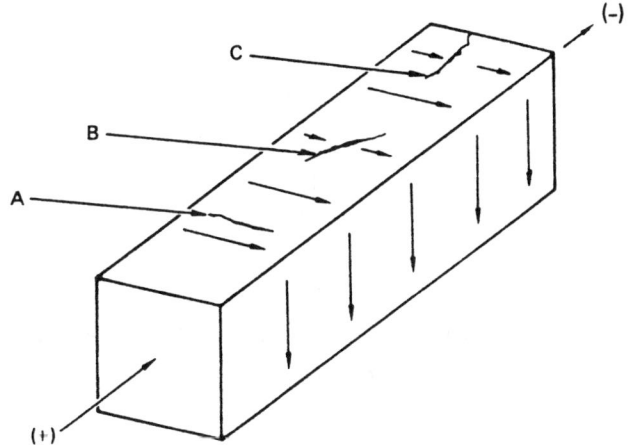

Crack A is parallel to the lines of force and will not have magnetic poles or flux leakage.

Cracks B and C will develop magnetic poles and flux leakage. Because of this, they will attract iron particles at the cracks.

Cracks that are crosswise (90° to the lines of force) cause more lines of force to be forced out at the surface giving a greater amount of flux leakage. Cracks can be up to 45° to the lines of force and still have enough flux leakage to adequately attract iron particles.

Perhaps by now you have noticed that cracks that lie in the same direction as the current flow (or at a maximum of 45°) are the ones that can be detected. Now let's see how this applies to you—the examiner/technician.

Turn ahead to page 2-22.

From page 2-18

The lines of force do NOT pass through the crack. Let's enlarge our view of the crack that lies parallel to the lines of force.

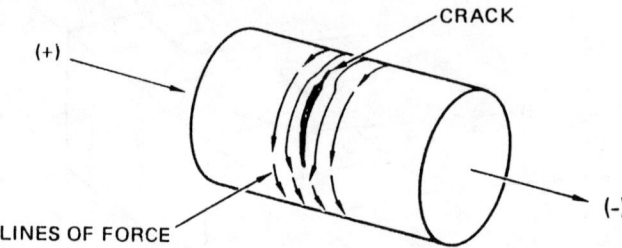

Notice how the lines of force *bend* around the crack. Since the lines of force are traveling in the same direction as the crack, there is very little area of the crack to force the lines of force out of the metal. The lines of force simply bend a little and remain in the metal.

Turn back to page 2-18 and try again.

From page 2-23                                                                 2-21

You think that crack B will NOT attract iron particles. Let's look at the steel bar again.

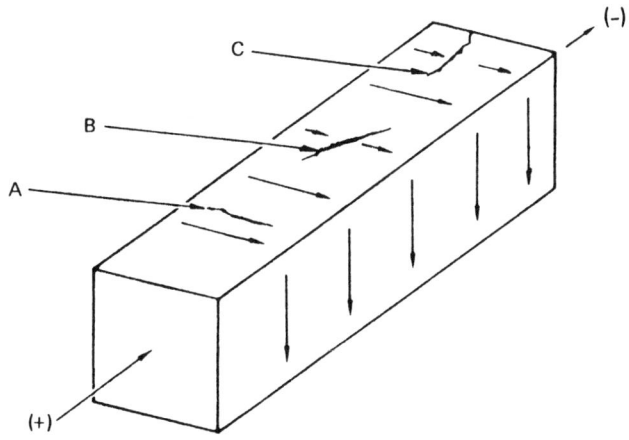

Remember that the magnetizing current runs from (+) to (−) left to right. Using the right-hand rule, the magnetic field is set up as shown by the arrows. Crack A is parallel to the lines of force (does not cut across any of the lines of force) so it will not form magnetic poles. Cracks B and C are at 90° and 45° to the lines of force and will form magnetic poles with flux leakage since they will disrupt the lines of force.

Return to page 2-23 and try once again.

*Circular magnetization* is accomplished in two ways.

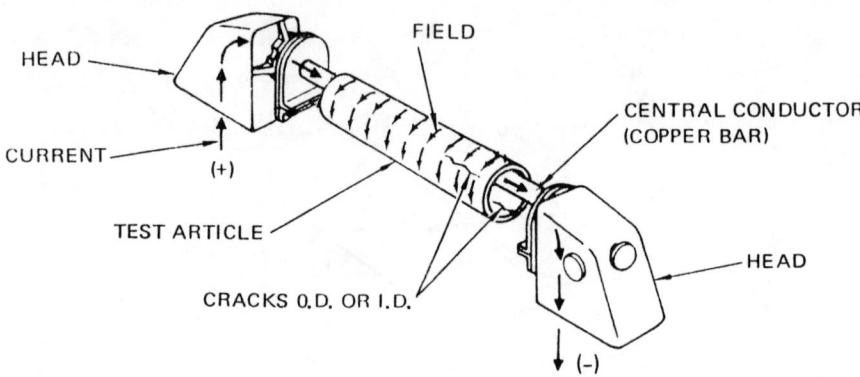

First, by passing electric current through a *central conductor* as in the illustration above; and second, by passing electric current through the test specimen itself as in the illustration below.

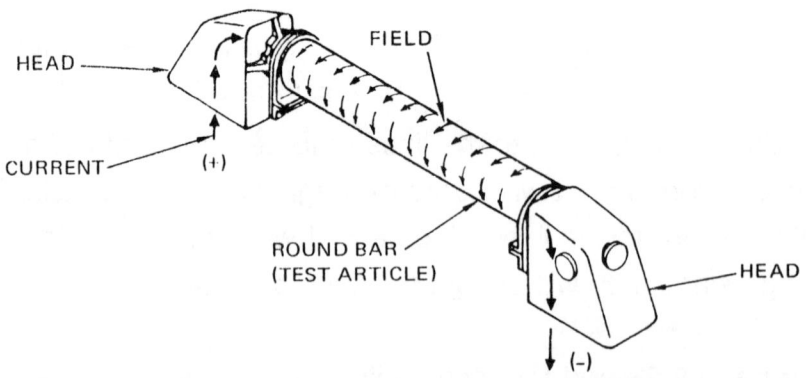

Passing electric current directly through the test specimen is called a *head shot* and causes a circular magnetic field within the specimen.

**If the round bar had a seam in it, do you think that iron particles would be attracted to the seam?**

**Yes** .................................................. **Page 2-27**
**No** ................................................... **Page 2-28**

From page 2-18

2-23

Correct. There will be no flux leakage if there are no north and south poles. Any cracks that are parallel with the lines of force will not attract iron particles.

As we discuss applications of the magnetic particle testing method later in the program, it will be apparent that test article and discontinuity orientation and shape will be important variables. Flux leakage will vary and be dependent upon these and other factors.

Here is another cracked steel bar.

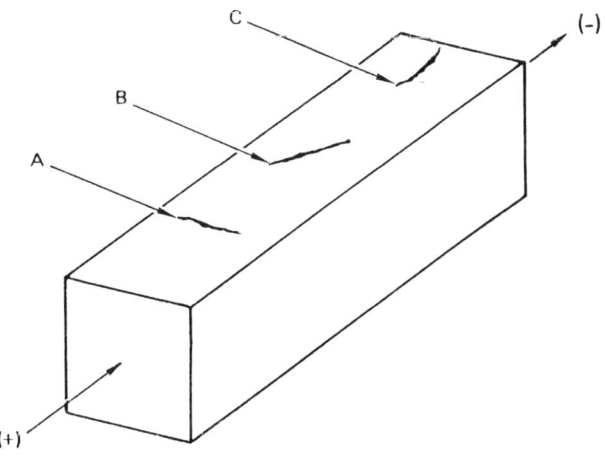

**If the steel bar is circularly magnetized, which of the cracks (A, B, or C) will *NOT* attract iron particles?**

A .................................................... **Page 2-19**
B .................................................... **Page 2-21**
C .................................................... **Page 2-25**

From page 2-18                                    2-24

You must be looking at the wrong crack. Let's take another look.

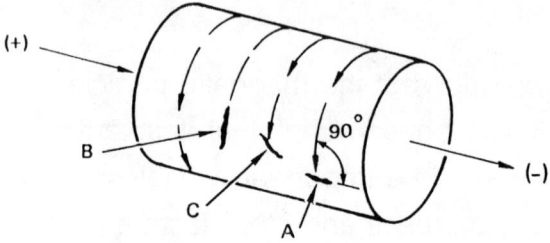

Notice that crack A cuts across (90° to) the lines of force. Crack B does not cut across any of the lines of force. Crack B runs in the same direction as the lines of force. In other words, crack B runs parallel to the lines of force.

Turn back to page 2-18 and select another answer.

From page 2-23                                                                 2-25

You think that crack C will NOT attract iron particles and that is incorrect. Let's look at the steel bar again.

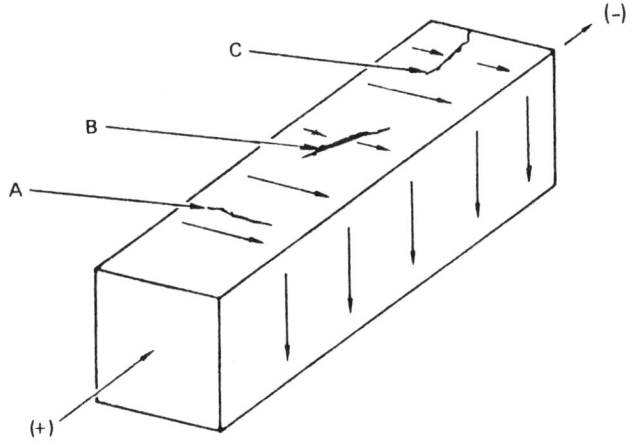

Remember that the magnetizing current runs from (+) to (-), left to right. Using the right-hand rule, the magnetic field is set up as shown by the arrows. Crack A is parallel to the lines of force (does not cut across any of the lines of force) so it will not form magnetic poles. Cracks B and C are at 90° and 45° to the lines of force and will form magnetic poles with flux leakage since they cut across the lines of force.

Turn back to page 2-23 and try another answer.

From page 2-29

We caught you napping that time.

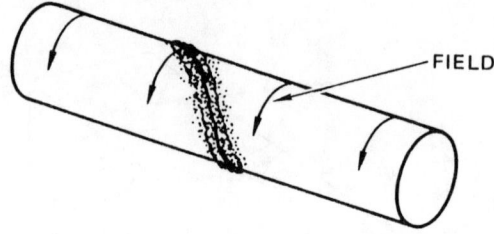

That shrink crack cuts across the lines of force at a 45° angle and *would* attract iron particles.

Turn ahead to page 2-31.

From page 2-22

2-27

Right. Iron particles would be attracted to the seam, and we would have an indication like this.

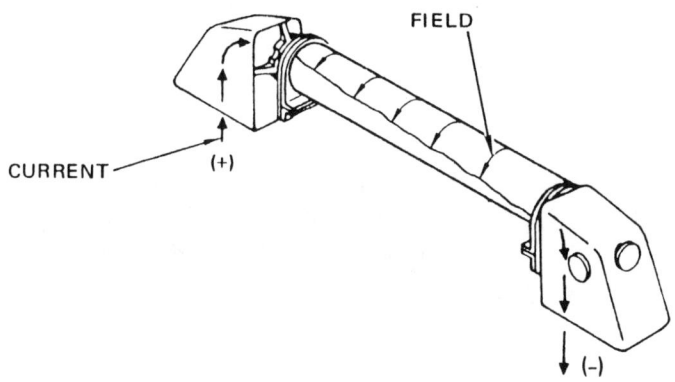

The seam is crosswise or transverse to the lines of force and would have flux leakage to attract the iron particles.

Suppose a round ferromagnetic steel bar were welded together in the middle like this.

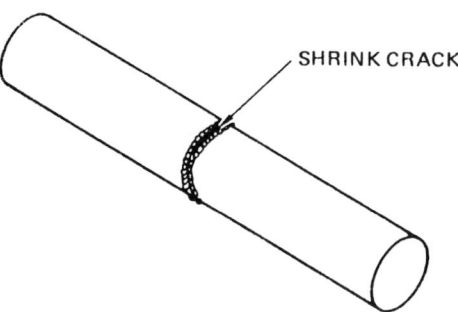

**Would this shrink crack in the weld attract iron particles if the bar were *circularly magnetized* between the heads?**

**No** .................................................. **Page 2-29**
**Yes** ................................................. **Page 2-30**

From page 2-22                                             2-28

You have probably forgotten what a seam is, so let's take a look at one.

Remember, a seam runs along the length of the bar. Apply the right-hand rule and you should be able to answer the question correctly.

Turn back to page 2-22 and try again.

Of course not. The shrink crack would NOT attract iron particles.

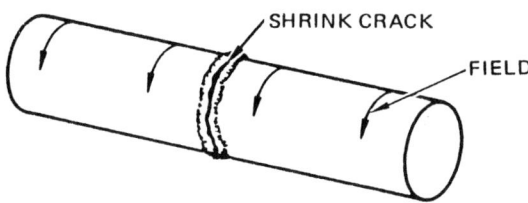

The shrink crack at the weld runs parallel to the lines of force and would NOT form flux leakage to attract iron particles.

Here is our round ferromagnetic steel bar with another weld in the middle at an angle of approximately 45° to the lines of force.

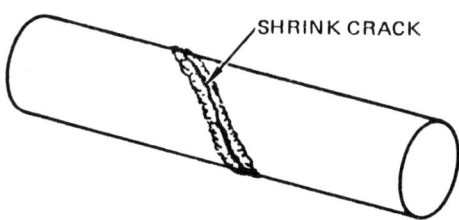

**If this part were circularly magnetized, do you think that iron particles would be attracted to a shrink crack in this weld?**

**No** .................................................. **Page 2-26**
**Yes** ................................................. **Page 2-31**

From page 2-27

You missed on this one, so let's review the facts in this case.

- The shrink crack is oriented around the circumference of the bar.

- By applying the right-hand rule, we know that the lines of force would also be oriented around the circumference of the bar.

- Therefore, the crack is in the same direction as the lines of force.

- Cracks that lie parallel to the lines of force will NOT form poles or flux leakage.

- Since NO flux leakage would be formed, iron particles would NOT be attracted to the crack.

Turn back to page 2-29.

From page 2-29                                                          2-31

Yes is right. Iron particles would be attracted to that shrink crack because it was approximately 45° to the lines of force.

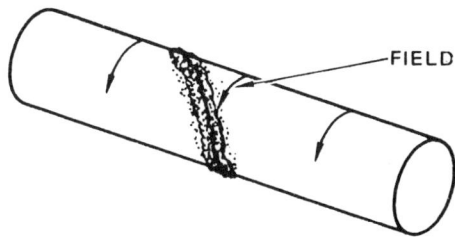

When a test specimen is magnetized between the heads, the magnetic field is strongest near the surface of the specimen. The magnetic field increases from zero at the center of the specimen to a maximum at the surface. The length of the specimen has little effect on the magnetic field established within it. If the specimen is uniform in all respects, the magnetic field will be uniform throughout its length.

Turn to the next page.

From page 2-31                                                    2-32

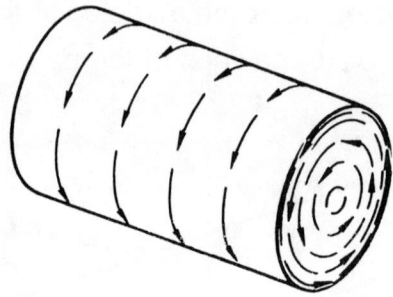

The strength of a magnetic field is often referred to as *flux density*. In this view, you can see that the lines of flux are crowded together near the surface. This shows that the flux density is greatest at the surface.

**The increased flux density would have what effect on flux leakage at a surface crack?**

**Decreased flux leakage** ............................ Page 2-34
**Increased flux leakage** ........................... Page 2-36

From page 2-36                                                           2-33

You selected point A. Flux density would NOT be greatest at point A. Point A is in the center and the magnetic field strength is zero there.

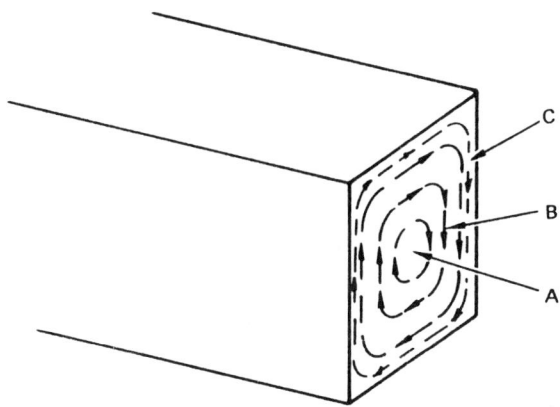

The strength of the magnetic field increases from zero at the center of the test specimen to a maximum near the surface.

Return to page 2-36 and select another answer.

From page 2-32

You feel that *increased* flux density would *decrease* flux leakage at a crack. No, that is not the case. Let us explain flux density more thoroughly.

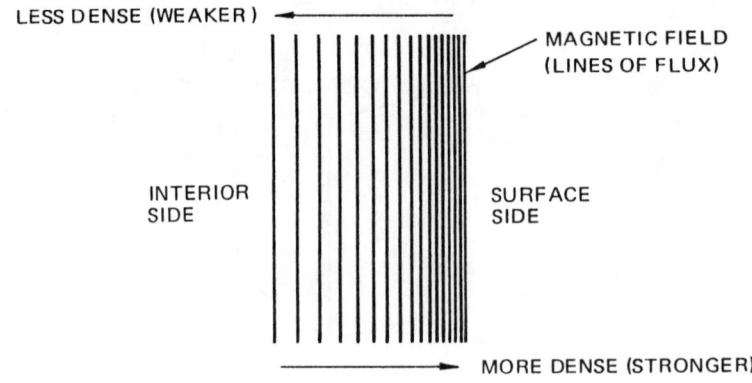

Lines of force are the same as lines of flux. Where these lines of flux are crowded together, the magnetic field will be the strongest. The strength of a magnetic field is known as its *flux density*. The more that lines of flux are crowded together, the greater will be the magnetic field flux density or strength.

With circular magnetization between the heads, the flux density (magnetic field strength) is greatest near the surface. Therefore, if the flux density is greatest near the surface, flux leakage would be greatest at a surface crack.

Turn ahead to page 2-36.

From page 2-39                                                                 2-35

You think that the flux density would remain the same just outside the surface of our round bar. Let's look at the diagram again.

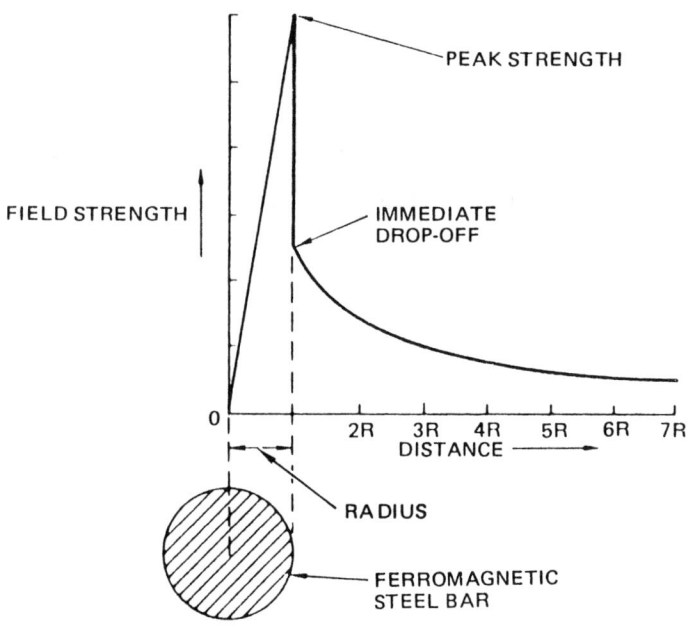

The vertical scale on the left of the chart is scaled to show the magnetic field strength or flux density. The scale across the bottom of the chart shows distance from the center of the bar. If the flux density remained the same just outside the surface of the bar, a line would have to be added at the peak strength point shown by the arrow. So you see, the flux density does NOT remain the same just outside the surface of the round bar.

Return to page 2-39 and study the problem again.

From page 2-32                                                              2-36

That's right.  The increased flux density would increase flux leakage at a crack.  Since the flux density is greatest near the surface, the magnetic field strength is also greatest in this area.

As a general rule it can be said that, with *circular magnetization* between the heads, flux density will be *greatest at the surface*.  This is particularly true with simple, uncomplicated parts such as the round steel bar.  However, trial and error methods must be used when magnetizing more geometrically complicated parts.

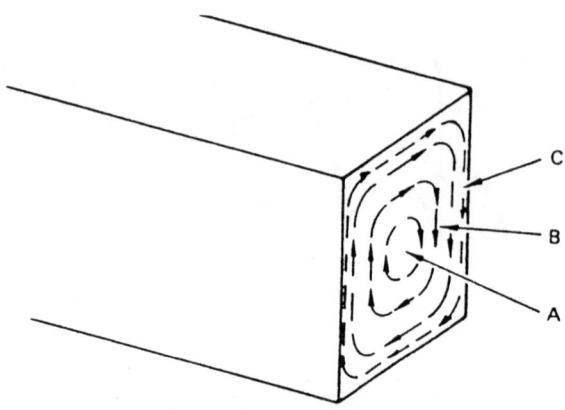

**If you were circularly magnetizing this ferromagnetic steel bar between the heads, at which point (A, B, or C) would you expect the flux density to be greatest?**

A . . . . . . . . . . . . . . . . . . . . . . . . . . . . . . . . . . . . . . . . . . . . . Page 2-33
B . . . . . . . . . . . . . . . . . . . . . . . . . . . . . . . . . . . . . . . . . . . . . Page 2-37
C . . . . . . . . . . . . . . . . . . . . . . . . . . . . . . . . . . . . . . . . . . . . . Page 2-38

From page 2-36

You selected point B. Flux density would NOT be greatest at point B.

The square steel bar is not one of those complicated articles that we mentioned. Magnetizing that bar between the heads would cause the magnetic field to be strongest at the surface. The strength of the magnetic field would increase from zero at the center of the test specimen to a maximum at the surface.

Return to page 2-36 and select another alternative.

From page 2-36

Excellent. Flux density would be greatest at point C. As a general rule, with circular magnetization between the heads, flux density will be greatest at the surface.

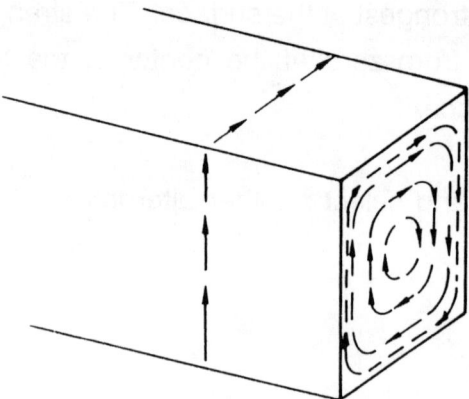

The magnetic field increases from zero at the center of the test specimen to a maximum at the surface. This is a general rule that applies to uncomplicated articles.

Turn to the next page.

| From page 2-38 | 2-39 |

Distribution of the magnetic field within an article being magnetized between the heads can be illustrated graphically. Here is what happens to our round, ferromagnetic steel bar.

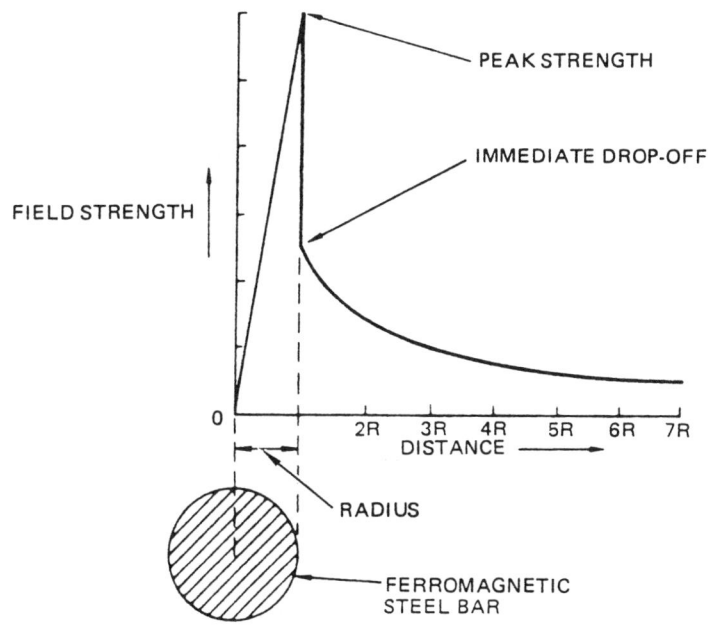

Here you can see that the field strength is zero at the center of the article. The flux density increases rapidly but evenly until it reaches peak strength at the surface.

**What happens to the flux density just outside the surface of our round steel bar?**

**Flux density remains the same** .................... **Page 2-35**
**Flux density increases** ........................... **Page 2-40**
**Flux density drops rapidly** ....................... **Page 2-43**

You think that the flux density increases just outside the surface of the round bar. You have misread the graph. Look at it again and see if you agree that the flux density peaks at the surface and *decreases* beyond the radius of the bar.

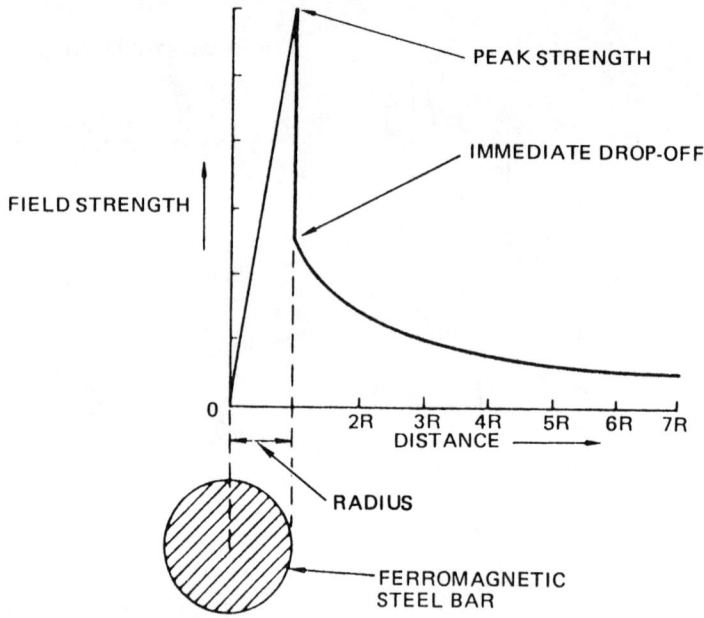

So you see, the flux density drops rapidly just outside the surface of the bar.

Turn ahead to page 2-43.

Let's expand the idea of *circular magnetization* using a *central conductor*. When electricity flows through an electrical conductor such as a copper wire, a magnetic field is established around the wire. In practice, we use this principle by putting a copper *bar* between the heads.

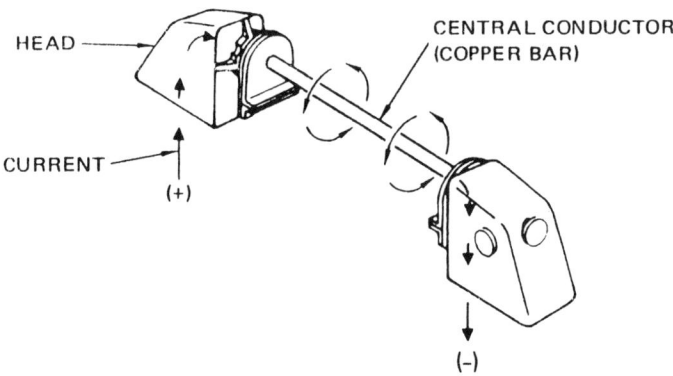

The copper bar is used as part of the equipment. When electricity flows through the copper bar, a magnetic field is established around the bar. The copper bar is called a *central conductor*.

The *central conductor* is used to establish a magnetic field in cylindrical objects, such as tubing and short hollow cylinders, and around holes through which the conductor can be placed. It is most effective when used this way because the magnetic field is *strongest at the surface of the central conductor*. In addition to the benefits received by inducing a magnetic field on the inner surface of a hollow part, the use of a central conductor eliminates the possibility of the part itself being damaged as a result of poor electrical contact with the heads.

Turn to the next page.

Let's take a look at the distribution of the magnetic field that is established in and around a solid copper bar when current is passed through it.

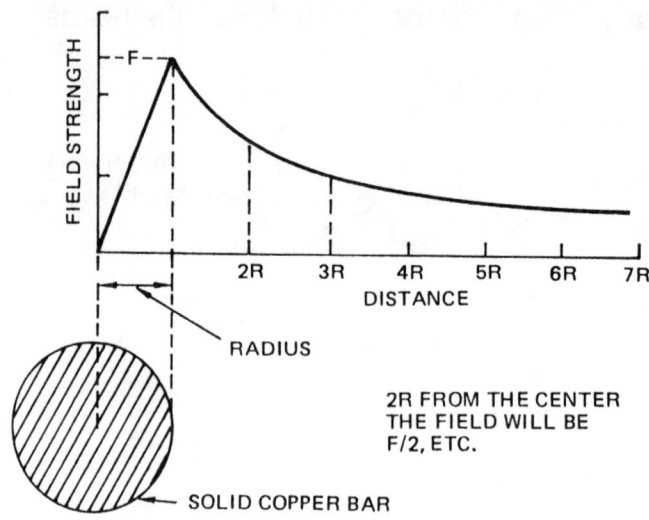

This illustration shows that the field strength (F) is zero at the center of the conductor and increases to the maximum at the surface of the conductor. Outside the conductor the field strength begins to decrease in proportion to the distance from the center of the conductor. At a distance of twice the radius of the conductor, the field strength is just half of the strength at the surface.

The important thing to remember is that *for a nonmagnetic conductor the field strength is greatest at the surface of the conductor.*

Turn ahead to page 2-44.

From page 2-39            2-43

That's right! Flux density drops rapidly just outside the surface of the round bar. However, at the surface of the article, flux density reaches its peak strength. This "peaking" of the flux density is the result of the magnetic material offering a lower resistance path than the surrounding air. The flux tends to stay in the material.

If we pass electric current through a *hollow* ferromagnetic steel bar as in magnetization *between the heads*, the magnetic field distribution in the bar can be shown graphically as follows.

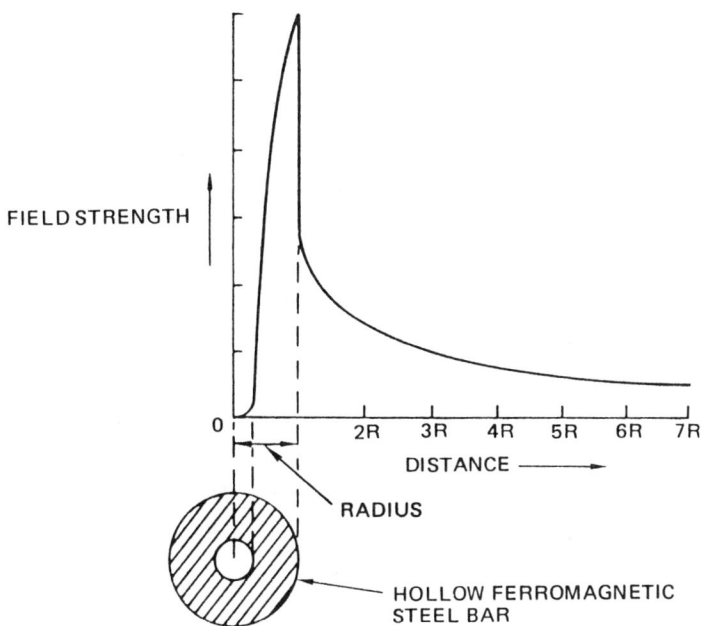

**When the hollow bar is magnetized between the heads, where is the area of *greatest* flux density?**

**In the center of the hole in the article** ............... **Page 2-45**
**At the outside surface of the article** ................ **Page 2-47**
**At the inner surface (ID) of the article** ............... **Page 2-49**

The magnetic field around the *central conductor* (bar of copper or aluminum) enters the cylindrical object and creates a circular magnetic field within the object. Since flux density is greatest at the surface of the central conductor, the strongest magnetic field will be induced in the article by allowing the article to lie on the central conductor. Unlike between-the-heads magnetizing, the use of the central conductor will create magnetic flux on the inner surface of the article as well as the outer surface. In fact, flux density is greatest on the inner surface and, depending on the wall thickness, something less on the outer surface.

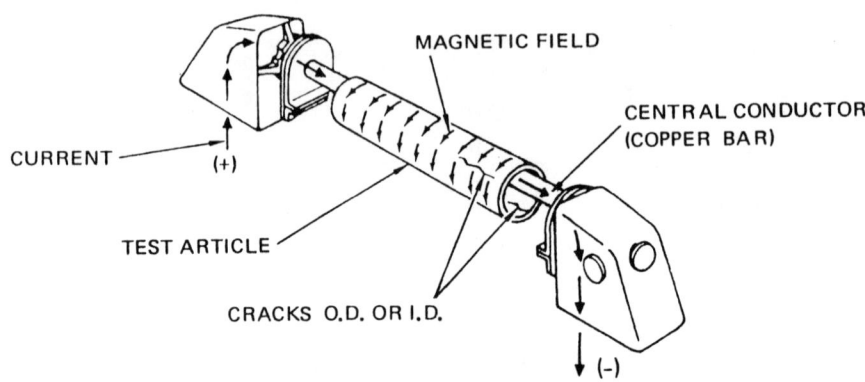

The circular magnetic field set up by the *central conductor* will detect cracks that are crosswise (45° to 90°) to the lines of flux as in the example above. These cracks cause flux leakage which attracts iron particles.

Turn ahead to page 2-46.

From page 2-43

You think that the area of greatest flux density in the hollow bar is the center of the hole in the bar. Actually, there would be NO magnetic field at the center. You may have been misled by that dashed line to the center of the hole. It was placed there to show only the radius of the article. Let's look at the chart again.

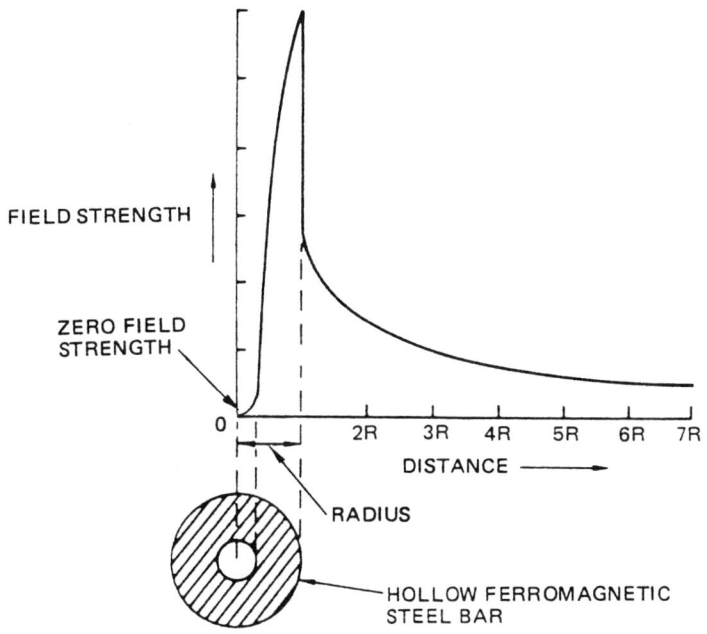

We have added an arrow here to show the point of zero field strength.

Turn back to page 2-43 and pick out the point of greatest flux density.

From page 2-44

**If we placed this ferromagnetic ring on a central conductor, would iron particles be attracted to a crack that runs parallel to the lines of flux like this one?**

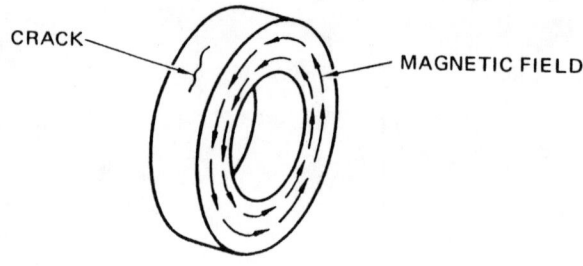

**Yes** .................................................. **Page 2-48**
**No** ................................................... **Page 2-50**

Absolutely. The flux density would be greatest at the outside surface of the hollow ferromagnetic steel article.

Remember that as a general rule, circular magnetization between the heads causes flux density to be greatest at the surface of the article. This is true with simple, uncomplicated articles. However, trial and error methods must be used when magnetizing more geometrically complex articles.

The point we are making here is that "between-the-heads" circular magnetization of a hollow article will result in strong magnetization at the outer surface. Magnetization at the inner surface will be minimal and discontinuities there may attract few, if any, iron particles.

Turn back to page 2-41.

You are probably confused because we are using the central conductor. But the rules don't change just because we changed the technique.

In this case, the crack is in the same direction as the lines of flux. As a result, no flux leakage was formed at the crack, so iron particles would not be attracted at the crack.

Turn ahead to page 2-50.

From page 2-43

You think that the area of greatest flux density would be at the inner surface (ID) of the hollow bar. No, the magnetic field distribution in the hollow bar is about the same as in the solid bar. The only difference is that the point of zero field strength in this example is at the center of the hole.

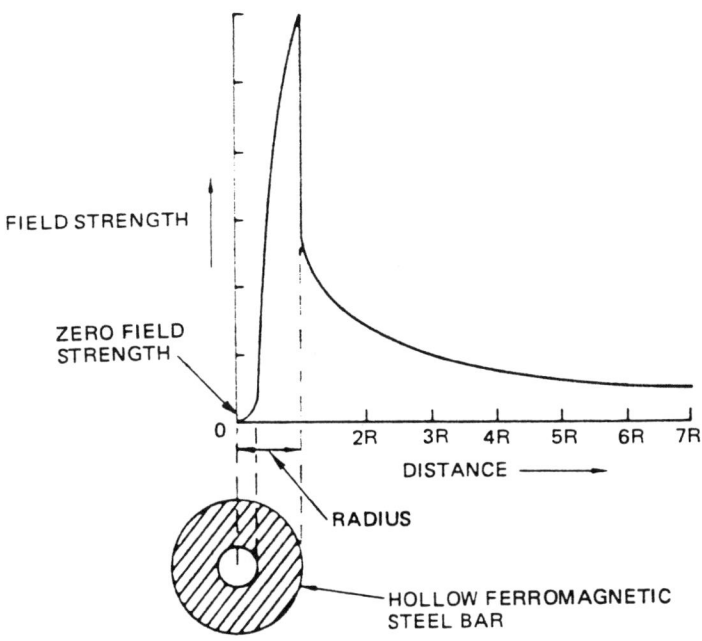

The arrow we have added, indicating zero field strength, should give you a clue to the correct answer.

Turn back to page 2-43 and try again.

No is correct. A crack that runs parallel to the lines of flux would not cause flux leakage and would not attract iron particles.

The *central conductor* is used to magnetize many different types of ferromagnetic hollow articles. Its greatest advantage is that the flux density is greatest at the surface of the bar and will induce a strong magnetic field which will locate cracks on both the inner and outer surfaces as illustrated here. Also, the use of a central conductor eliminates the possibility of the part itself being damaged as the result of poor contact with the heads.

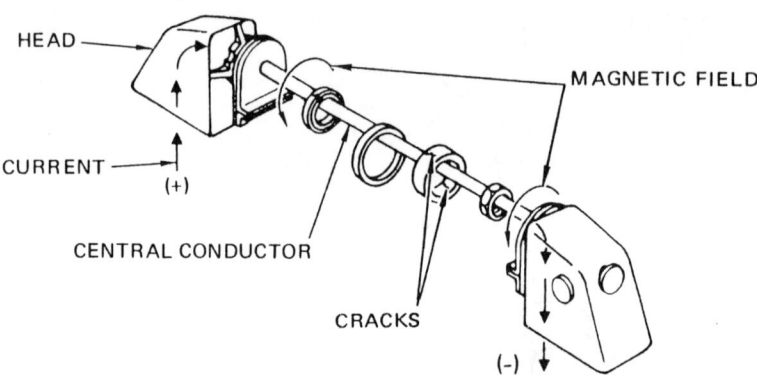

Turn to the next page.

From page 2-50

2-51

**In the example below, do you think circular magnetization with the CENTRAL CONDUCTOR would attract iron particles to the cracks shown in this gear?**

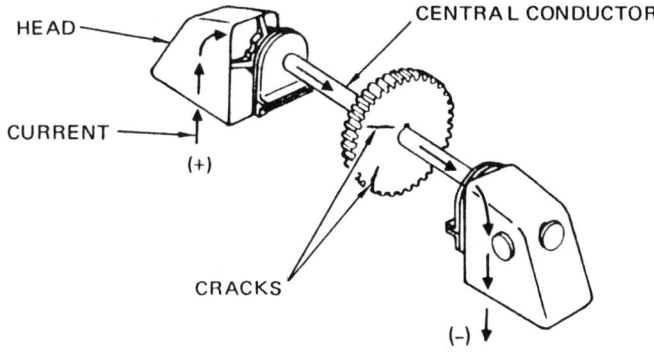

**Yes** .................................... **Page 2-53**
**No** ..................................... **Page 2-54**

## Longitudinal Magnetic Fields

In a longitudinal magnetic field, an article is magnetized lengthwise. The bar magnet is a good example of a longitudinal magnetic field.

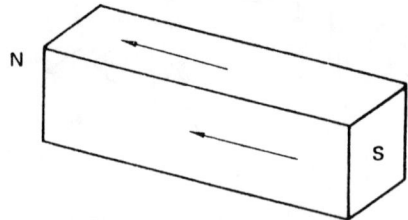

The magnetic lines of flux in the bar magnet go through the length of the bar. You will recall that a crack which runs across 90° to (or at least 45° to) the lines of flux will cause flux leakage. The flux leakage will attract iron particles like this.

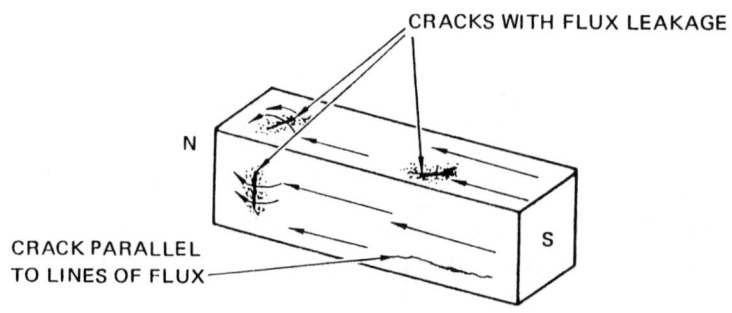

As you can see, a crack that runs parallel to the lines of flux will not cause flux leakage. Now let us see how an article can be longitudinally magnetized.

Turn ahead to page 2-55.

Right. Iron particles would be attracted to the cracks in the gear.

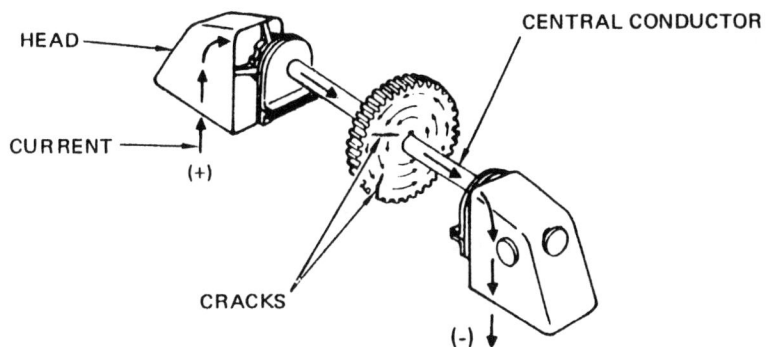

The cracks run crosswise (90°) to the lines of flux and would cause flux leakage. If we dust iron particles on the gear, they would be attracted to those cracks.

It is important to remember that flux density is greatest at the surface of a central conductor. Therefore, hollow articles should be placed on the central conductor as shown above and allowed to come in direct contact with it to obtain the strongest magnetic field.

The diameter of the central conductor should be as large as is practical for the situation. The magnetic field is effective only for a distance along the circumference of the article equal to four times the diameter of the central conductor. This means that large articles will have to be rotated on the conductor to inspect the entire circumference.

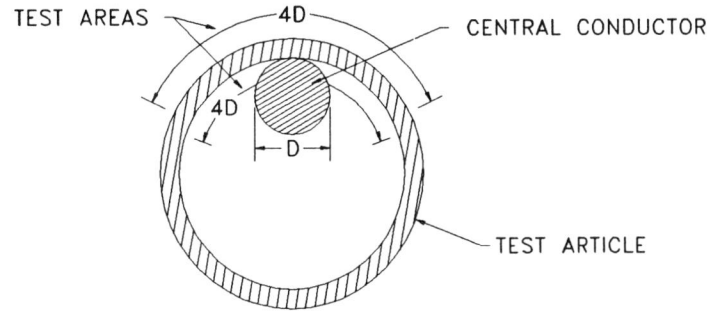

Turn back to page 2-52.

From page 2-51                                                                  2-54

Use of the central conductor is a little tricky, so let's look at the lines of flux set up in the gear.

Up to now, we have been looking only at the outer surface of bars. With use of the central conductor, we can now look at the ends or flat surfaces of the articles. The circular field is present in the flat end surfaces of the gear as well as in the outer rim.

Take another look at the illustration on page 2-51 and try the question again.

Longitudinal magnetization also uses the principle that electric current passing through a copper wire forms a magnetic field around the wire.

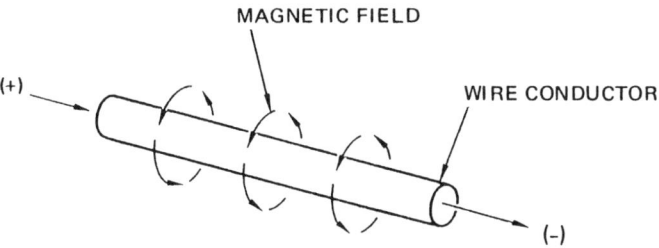

When the copper wire is wound into a coil, the lines of flux around each turn of the coil combine with those of each of the other turns in the coil. This increases the flux density and gives a total force in a longitudinal direction.

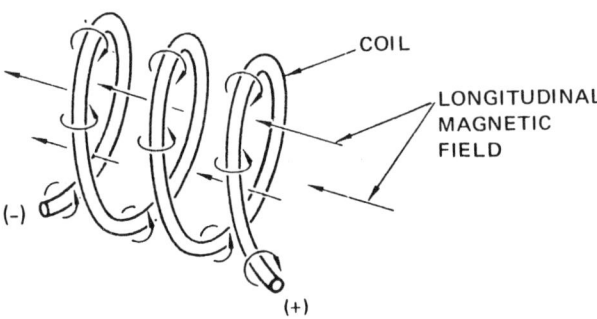

The flux density or strength of the magnetic field is greatest at the surface of the copper wire. Therefore, flux density of the total longitudinal magnetic field will be greatest *at the inside surface of the coil.*

Turn to the next page.

The field strength developed *within* a coil depends on three factors:

- The number of turns in the coil.

- The current flowing through the coil.

- The diameter of the coil.

Obviously, the magnetic field of one turn is increased by the presence of a second turn and, just as obviously, the field strength would increase as a higher current is passed through the coil. Not so obvious is the effect of a change in the diameter of the coil.

Consider the following: If two coils with different sized openings have the same number of turns and are energized with the same amount of current, they each will generate the same amount of flux, right? Right.

But in the coil with the smaller opening the lines of force are forced closer together. The flux density (flux per unit area) is greater inside the small coil than inside the larger coil.

However, don't be misled. Although the field within the coil opening is stronger per unit area, for all practical purposes the field at the inside surface of the coil is the same for both coils.

Turn to the next page.

When we place an article inside the coil through which electric current is passing, a *longitudinal magnetic field* is set up in the article.

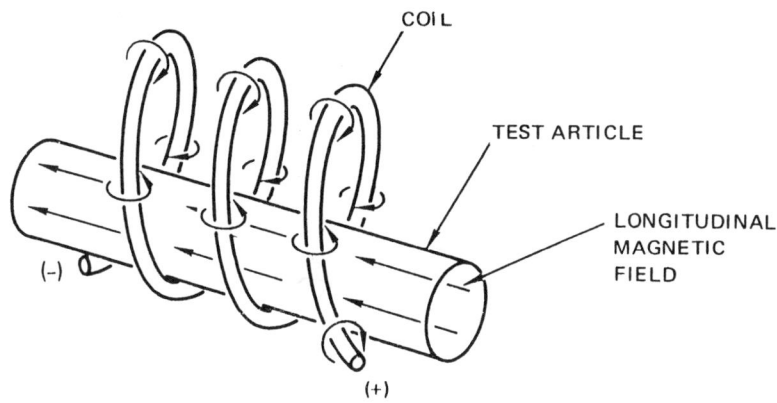

The longitudinal magnetic field will cause flux leakage at cracks which run crosswise to the lines of flux. Cracks running up to 45° to the lines of flux will also have flux leakage.

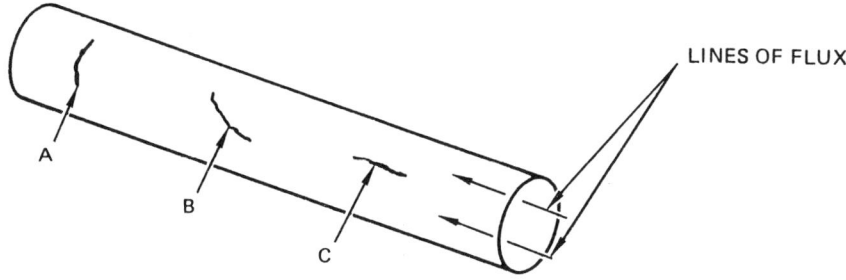

**With this round bar longitudinally magnetized, which of the cracks will attract iron particles?**

| | |
|---|---|
| A | Page 2-58 |
| A and B | Page 2-61 |
| B and C | Page 2-62 |

From page 2-57

Yes, crack A will attract iron particles. It runs crosswise (90°) to the lines of flux and will have flux leakage.

One thing more to remember is that cracks up to 45° to the lines of flux will also have magnetic poles and/or flux leakage.

Turn back to page 2-57 and see if there isn't a more complete answer.

From page 2-61

In practice, we use a coil similar to the one we have been using as an example to produce a longitudinal magnetic field.

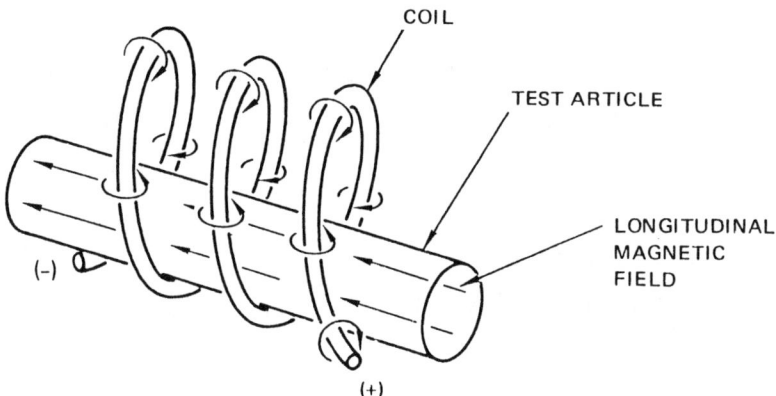

However, the coils are pushed together and placed inside a housing.

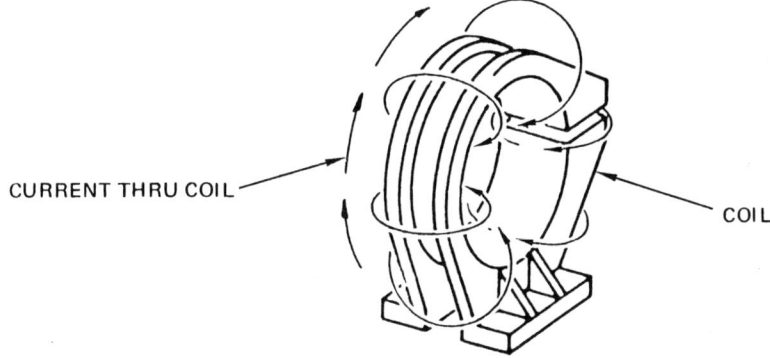

Turn to the next page.

From page 2-59                                                              2-60

The magnetic field is strongest near the *inside surface* of the coil where the flux density is greatest. Flux density decreases toward the center of the coil where it is zero.

**If you were to longitudinally magnetize a round, ferromagnetic steel bar in the coil, where would you place the bar to get the greatest flux density?**

**Near the center of the coil** ......................... **Page 2-63**
**Near the inside surface of the coil** ................. **Page 2-64**

Right. A and B was the best selection.

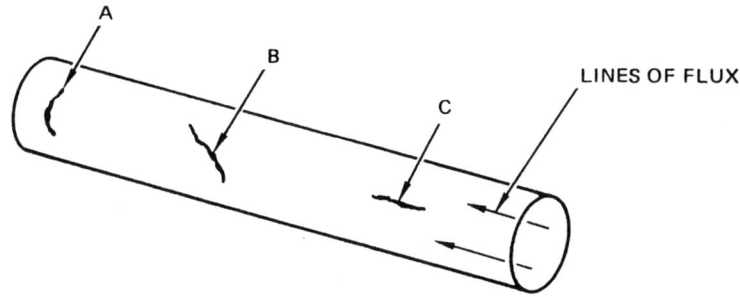

Crack A is crosswise (90°) to the direction of the lines of flux and would have flux leakage to attract iron particles to the crack. Crack B is about 45° to the lines of flux and would also have flux leakage to attract iron particles.

Crack C is parallel to the lines of flux and would not disrupt the lines of flux and cause flux leakage. The lines of flux follow the path of least resistance and squeeze around crack C staying in the metal.

Now, let us see how the longitudinal magnetic field is used.

Turn back to page 2-59.

From page 2-57                                                              2-62

You are half right. Crack B will attract iron particles because it is 45° to the lines of flux.

Crack C will not attract iron particles. It is parallel to the lines of flux and will not disrupt them.

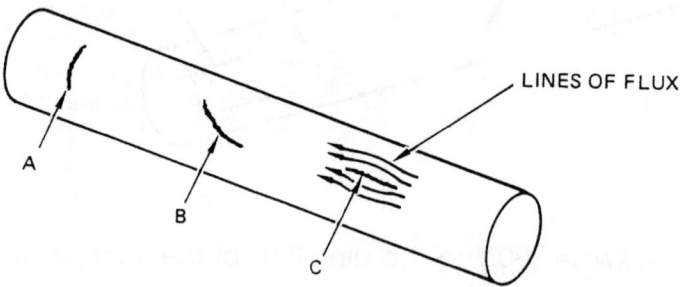

The lines of flux tend to follow the path of least resistance and that is to stay in the metal. Therefore, the lines of flux squeeze around crack C and do not cause flux leakage.

Turn back to page 2-57 and pick a better answer.

You missed the point.  Let's look at that coil again.

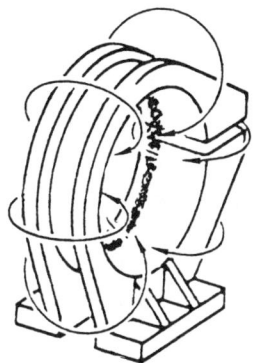

The magnetic field is strongest near the *inside surface* of the coil as shown above by the arrows surrounding the coil. The shaded area shows where the field is strongest.

The point to be made here is that the magnetic lines of flux are concentrated near the inside surface of the coil. This concentration of lines of flux causes the magnetic field to be strongest near the inside surface of the coil. Toward the center of the coil, the lines of flux are not so close together. At the center of the coil, the magnetic field decreases to zero.

To longitudinally magnetize a round steel bar in the coil, you would place the bar where the magnetic field is strongest—the area of greatest flux density near the inside surface of the coil.

Turn to the next page.

From page 2-60

Right. The magnetic field is strongest near the inside surface of the coil where the flux density is greatest. That is where the lines of flux are concentrated.

A word of caution: if a ferromagnetic article is not held against the inside surface of the coil when the current is turned on, it may be pulled violently up next to the coil. This can be surprising if unexpected.

If the article is not attracted to the coil, it is *not* ferromagnetic material.

A longitudinal magnetic field can also be used to locate cracks in hollow, tube-like articles. The cracks must be crosswise, or at least 45° to the lines of flux, to attract iron particles to the crack.

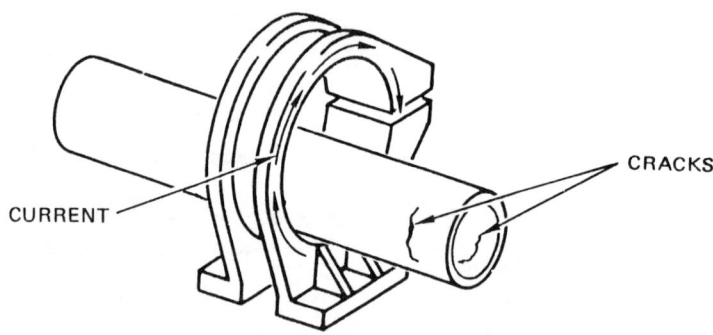

**If we shoot electricity through the coil (coil shot) to longitudinally magnetize this tube, do you think that the crack on the outside and the crack on the inside would both attract iron particles?**

**Yes** .................................................. Page 2-66
**No** ................................................... Page 2-68

From page 2-66                                                          2-65

That is a tough one to lose, but iron particles would NOT be attracted at the lamination. Perhaps the lamination on the right end of the bar threw you, since it could appear to be crosswise to the lines of flux. The problem here is that you are only seeing the end and side of the lamination. Let's enlarge the view and take another look at it.

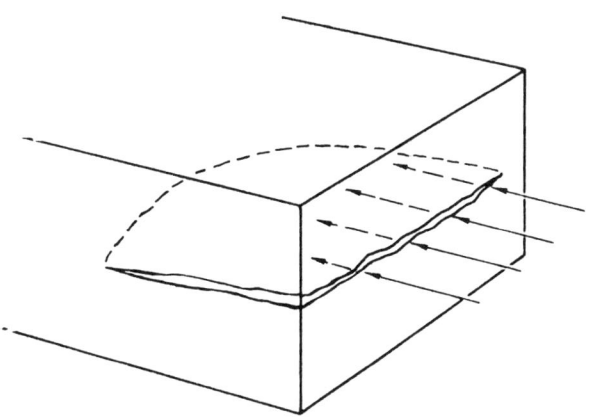

Notice that the lamination extends into the metal in the same direction as the lines of flux. The lamination does not cut across the lines of flux, so iron particles would not be attracted to the lamination.

Turn ahead to page 2-69.

That's right. Both cracks in the tubing would attract iron particles since they are crosswise (90°) to the lines of flux.

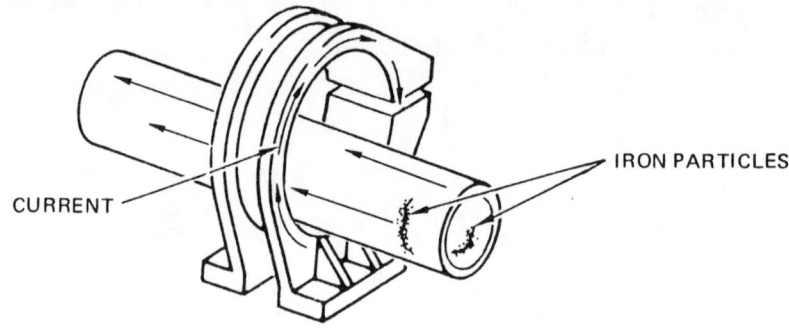

Below, a piece of iron bar is being longitudinally magnetized.

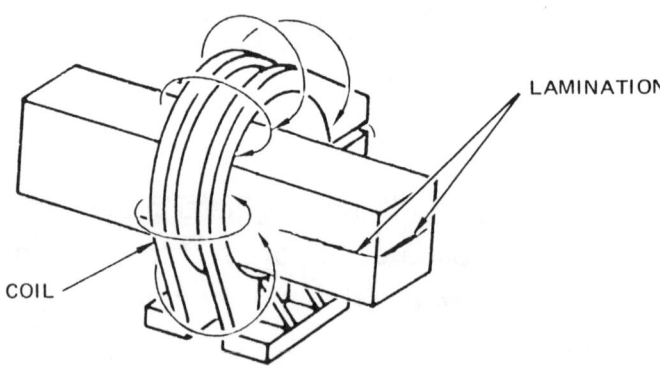

**Do you think that iron particles would be attracted to the lamination if the article were longitudinally magnetized?**

**Yes** ............................................. **Page 2-65**
**No** .............................................. **Page 2-69**

Right. It would take two coil shots to adequately magnetize an article 20 inches (51 cm) long. Any article under 12 to 18 inches (30 to 46 cm) long would only require one coil shot.

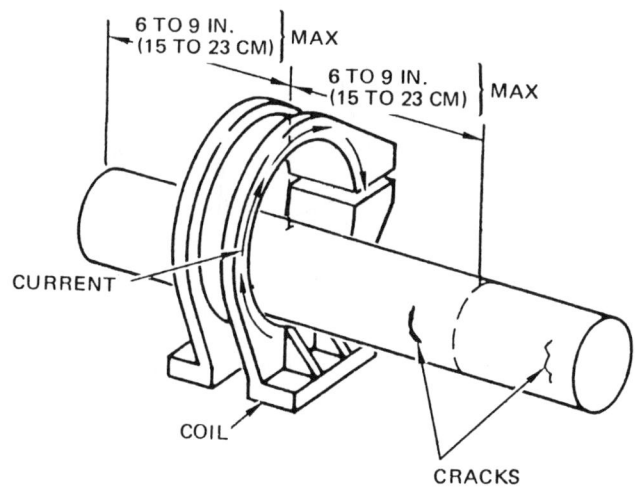

To attract iron particles to the crack on the right, the article would have to be moved to the left so that the crack would be within 6 to 9 inches (15 to 23 cm) of the edge of the coil.

The 6 to 9 inch (15 to 23 cm) rule-of-thumb is based on the amount of current used and the permeability of the material being magnetized. Effective use of the rule must be based on experience with its application.

Turn ahead to page 2-71.

You must have forgotten that any crack which cuts across the lines of flux will cause flux leakage. Let's add the longitudinal lines of flux to the article and see what we have.

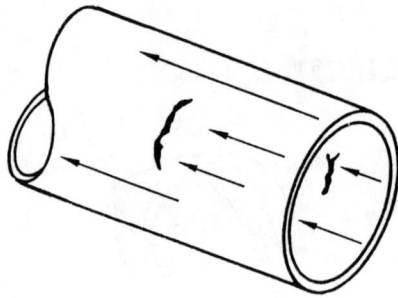

Here is an enlarged view of a portion of the longitudinally magnetized tube. The lines of flux are present on the inside surface of the tube as well as the outside.

Since the cracks cut across the lines of flux, iron particles will be attracted to both cracks.

Turn back to page 2-66.

From page 2-66                                                                 2-69

"No" is the correct answer. Iron particles would not be attracted to the lamination.

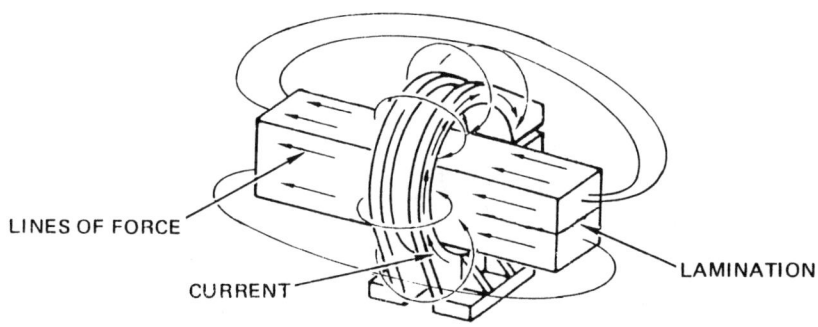

The laminations are oriented in the same direction as the lines of flux; therefore, there would be no flux leakage to attract iron particles.

The effective length of the magnetic field in an article magnetized with a coil is 6 to 9 inches (15 to 23 centimeters) *on either side of the coil*. The 6-inch to 9-inch rule is a variable resulting from the differences in permeability of the various ferromagnetic materials. For example, the effective length of the field for soft iron which is highly permeable would probably be 9 inches (23 cm). The effective length for hard, ferromagnetic steel which has low permeability might be 6 inches (15 cm).

Turn to the next page.

From page 2-69

Any cracks within the 6-inch (15 cm) to 9-inch (23 cm) range *on either side* of the coil will develop sufficient flux leakage to attract iron particles.

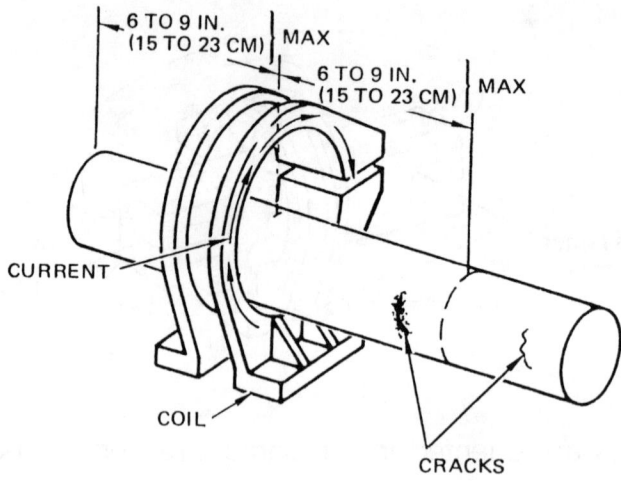

Cracks that are beyond the 6-inch to 9-inch range will not have sufficient flux leakage. In other words, an article longer than 12 to 18 inches (30 to 46 cm) would require two coil shots.

**How many coil shots would it take to adequately magnetize an article 20 inches (51 cm) long?**

**Two** . . . . . . . . . . . . . . . . . . . . . . . . . . . . . . . . . . . . . . . . . **Page 2-67**
**One** . . . . . . . . . . . . . . . . . . . . . . . . . . . . . . . . . . . . . . . . . **Page 2-73**

From page 2-67                                                          2-71

## Magnetization by Cable

Sometimes test articles are too big to fit into the ordinary coil. When this happens, an insulated copper cable can be used to form a coil for *longitudinal magnetization* of the article. Here is an example.

When the cable is wrapped around the object to be magnetized, electric current passing through the cable creates a longitudinal magnetic field. The effective distance of the longitudinal magnetic field created by the cable is the same as the effective distance of a stationary coil.

**Which of the following is the correct effective distance of the magnetic field?**

**6 inches (15 cm) on both sides of cable** . . . . . . . . . . . . . **Page 2-72**
**6 to 9 inches (15 to 23 cm) on both sides of cable** . . . . . **Page 2-75**
**12 to 18 inches (30 to 46 cm) on both sides of cable** . . . **Page 2-76**

From page 2-71

Yes, the magnetic field is effective for 6 inches (15 cm) on either side of the coil, but that is not all there is to it.

The permeability of the material is the deciding factor. For example, soft iron and iron with a low carbon content are highly permeable, and the effective distance of the longitudinal magnetic field may run as high as 9 inches (23 cm) on *both* sides of the coil. On the other hand, with hard steel of high carbon content, the effective magnetic field may be as low as 6 inches (15 cm).

Turn back to page 2-71 and try again.

One coil shot would not adequately magnetize an article 20 inches (51 cm) long. Remember, the maximum effective distance of a coil shot is 18 inches (46 cm) with easily magnetized (highly permeable) material. Some kinds of material which have low permeability may require as little as 12 inches (30 cm) per shot. In any event, an article that is over 18 inches (46 cm) long will require two coil shots.

The key words in the statement above are "adequately" and "effective." These words tell you that a field does exist beyond 9 inches (23 cm) from the coil but it is in no way adequate to create effective fields. The field from 9 to 18 inches (23 to 46 cm) from the coil is strong enough, however, to magnetize any ferromagnetic material inadvertently moved to within this range. Any parts remaining more than 18 inches (46 cm) from the coil will not be magnetized adequately for inspection purposes.

Turn back to page 2-67.

From page 2-75

2-74

## Use of Prods

Prods are *current-carrying conductors* (round copper bars) which are used to magnetize localized areas.

### Caution

The use of prods may be restricted for many applications due to the possibility of electrical arc burns at the points of contact with the test article. The test operator should determine the acceptability of prod use before proceeding with the test.

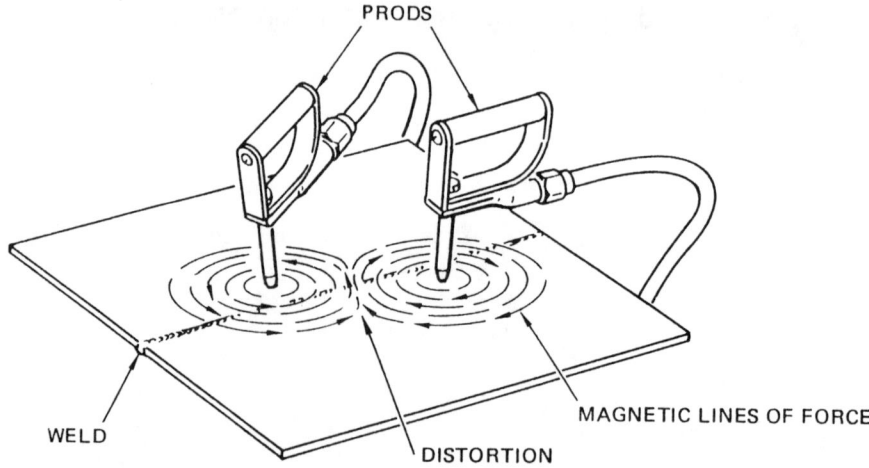

Prods are connected by cable to the current source. When electric current flows through the prods, a distorted circular magnetic field is created in the test article.

**Let's imagine a straight line drawn from one prod to the other. In your judgment, the field is:**

**parallel to the imaginary line** . . . . . . . . . . . . . . . . . . . . . . . **Page 2-78**
**perpendicular to the imaginary line** . . . . . . . . . . . . . . . . . **Page 2-80**

Correct. The effective distance of the longitudinal magnetic field created by the cable coil is 6 to 9 inches (15 to 23 cm) on both sides of the cable.

Above is another example using an insulated copper cable to create a longitudinal magnetic field in an article. In this case, the cable is connected to the heads for a source of electric current.

Turn back to page 2-74.

From page 2-71

2-76

No, "12 to 18 inches (30 to 46 cm) on both sides of the cable" is NOT the correct answer. You are confusing the total distance with the distance on both sides of the cable. In other words, the total effective distance of the longitudinal magnetic field within the coil is 12 to 18 inches (30 to 46 cm).

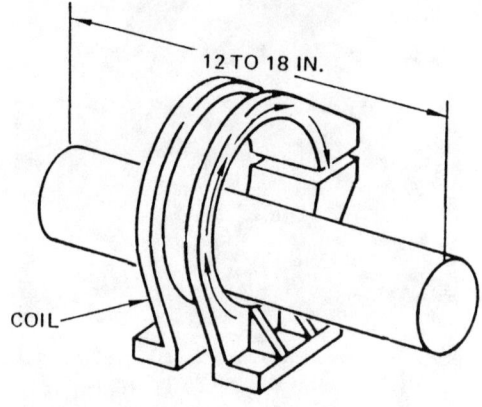

Turn back to page 2-71 and try again.

From page 2-80                                                              2-77

Yes, of course. A longitudinal shrink crack in the weld between the prods would attract iron particles. A longitudinal shrink crack would be nearly 90° to the lines of flux. This would cause flux leakage which would attract iron particles.

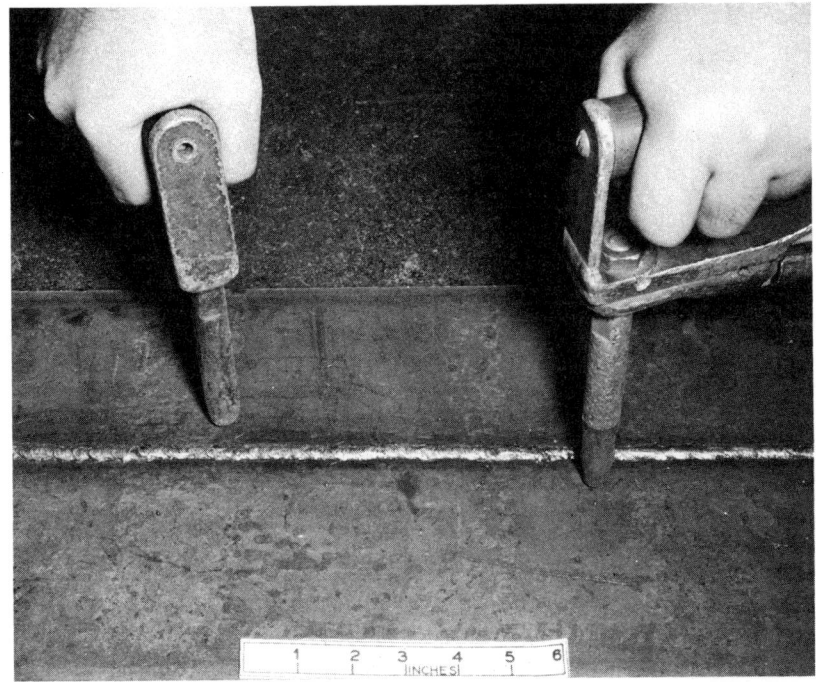

Prod magnetization is most effective when the prods are spaced 6 to 8 inches (18 to 20 cm) apart and in line with the suspected discontinuity as in the above picture.

Turn ahead to page 2-81.

You are probably confused by "imaginary lines." Here's the same illustration with the imaginary line drawn in.

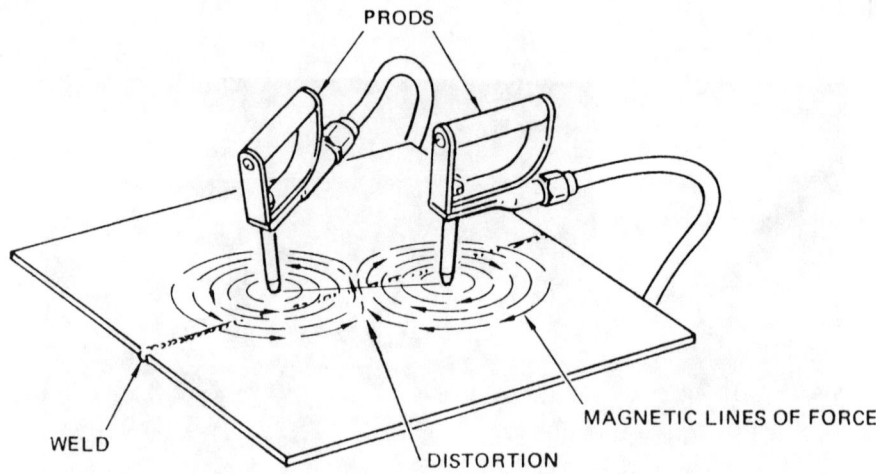

Now, note that even though the field is circular, the magnetic lines of force cross at 90° to the line drawn between the prods. We call this to your attention since it is along the line between the prods that the strongest indications will be obtained.

Now turn ahead to page 2-80.

From page 2-80 | 2-79

You selected "No." Well, let's take another look at this and see if you still feel that way.

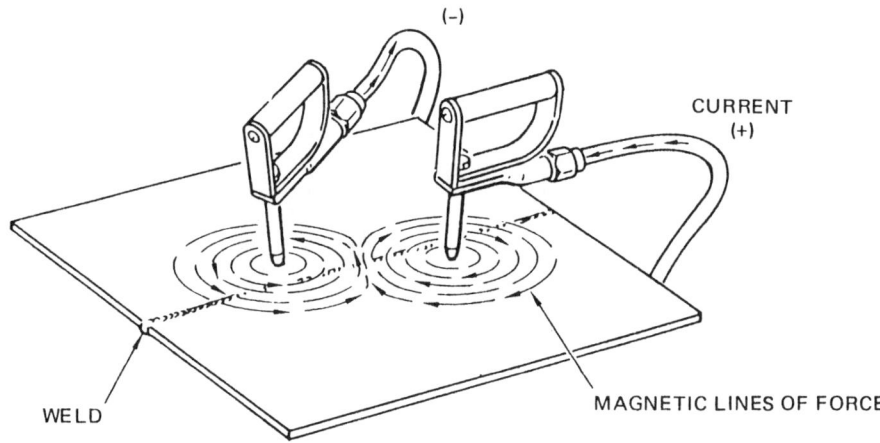

The rule that cracks which are between 45° and 90° with the lines of flux will cause flux leakage also applies to use of prods. In this case, the shrink crack in the weld is located between the prods. The lines of flux between the prods are crossing the weld between 45° and 90°, so iron particles would be attracted to the shrink crack.

Turn back to page 2-80 and study the problem again.

That's right! The field is perpendicular to a line drawn between the two prods. It is essentially along this line that the current flows.

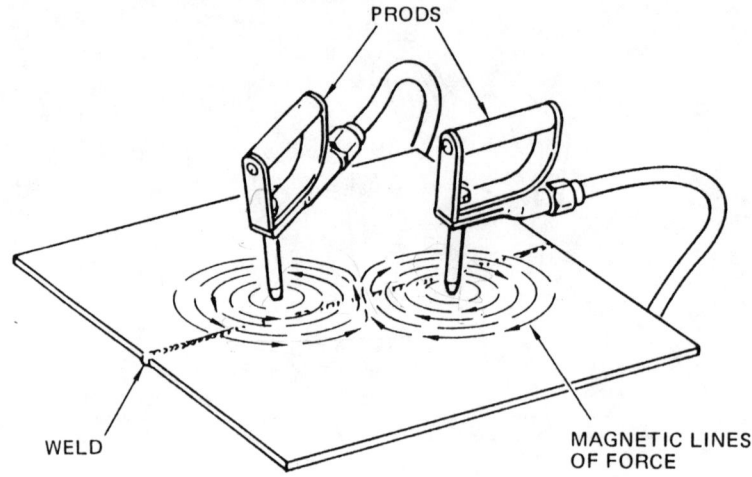

Note also that the flux density is greatest along this same line.

**If a longitudinal shrink crack were located in the weld between the prods, would iron particles be attracted to the crack?**

Yes .............................................. Page 2-77
No ............................................... Page 2-79

## Use of a Yoke

A *yoke* is a U-shaped piece of ferromagnetic metal with a coil wound around it to carry the magnetizing current. When the coil is energized while a specimen is placed across the poles of the yoke, a *longitudinal magnetic field* is set up in the test specimen. In contrast to the prods, the *yoke* actually applies the magnetic field directly to the test article. In this way, cracks that have their major axis perpendicular to an imaginary line between the yoke 'legs' will produce the greatest amount of flux leakage.

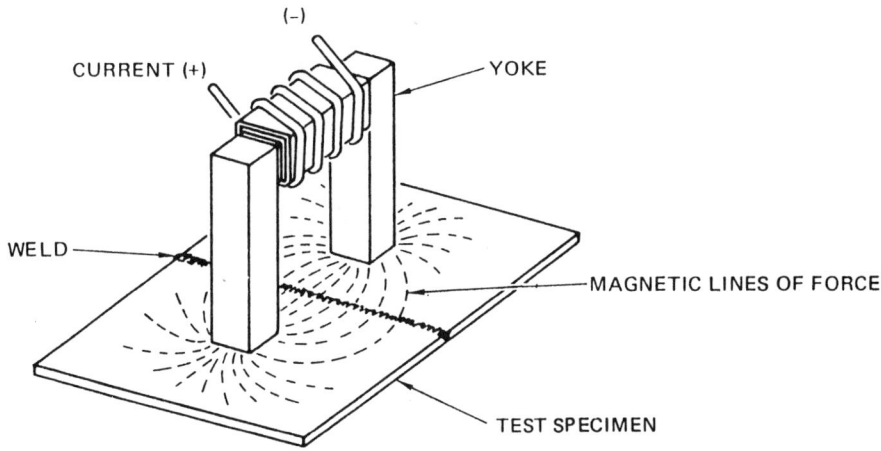

The magnetic field produced by the yoke does not lie entirely within the specimen, and an external field is produced that can attract iron particles. This external field is a deterrent to locating discontinuities lying beneath the surface. However, if the iron particles are applied sparingly and directed at the area between the poles, indications of surface discontinuities are easily seen.

Now turn to the next page for a chapter review.

## CHAPTER REVIEW

_A_  1.  The central conductor method is used to circularly magnetize different types of _____ articles such as tubes, rings, and nuts.

  A. hollow
  B. round
  C. square
  D. small

_A_  2.  The current through a wire always flows from _____ to negative.

  A. positive
  B. left
  C. negative
  D. longitudinal

_C_  3.  Suppose that we suspected that there might be a seam in this bar. To find out, we would _____ the bar.

  A. demagnetize
  B. longitudinally magnetize
  C. circularly magnetize
  D. prod

From page 2-82

__D__  4.  Because flux density is greatest at the surface of the central conductor, a hollow article will have greatest flux density at the _____ surface of the article.

    A.    bottom
    B.    outside
    C.    top
    D.    inside (inner)

__B__  5.  The magnetic field produced by the current flowing through a wire is in a _____ direction.

    A.    perpendicular
    B.    circular
    C.    parallel
    D.    oblong

__C__  6.  The magnetic field set up by a yoke is essentially:

    A.    circular.
    B.    useless.
    C.    longitudinal.
    D.    surface following.

__A__  7.  The flux density of the magnetic field around a ferromagnetic steel bar drops rapidly just _____ the surface of the bar.

    A.    outside
    B.    inside

From page 2-83                                                      2-84

D   8.  Whenever current is passed through any magnetic article, a circular field is produced:

A. only inside it.
B. only outside it.
C. just beyond its surface.
D. inside and around it.

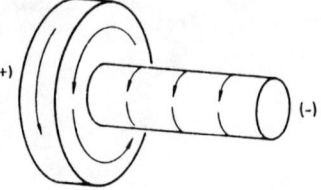

A   9.  A crack that runs crosswise to the lines of force will cause flux leakage which _____ iron particles.

A. attracts
B. repels
C. absorbs
D. discolors

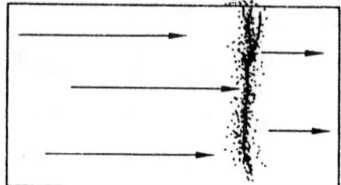

B   10. Here is a bar with a seam in it. If the bar were circularly magnetized, the seam would run _____ to the lines of force.

A. parallel
B. crosswise
C. 45°
D. 60°

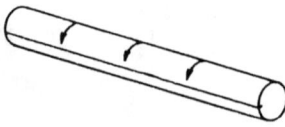

From page 2-84

__D__ 11. Any crack that runs from _____ to 90° across the magnetic field will form flux leakage.

A. 15°
B. 25°
C. 35°
D. 45°

__A__ 12. If we suspected that this weld had a circumferential shrink crack in it, would a circular field cause flux leakage at this crack?

A. No
B. Yes
C. Can't be determined

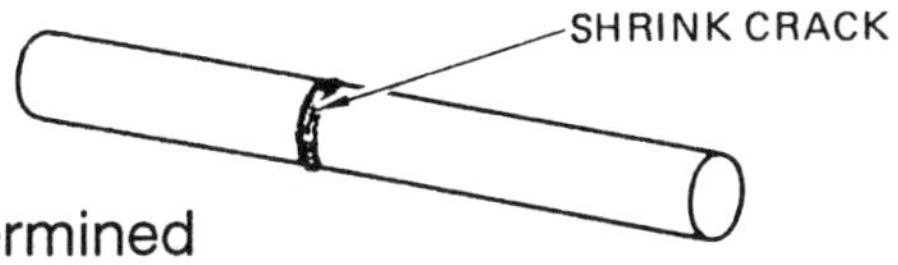

__C__ 13. A circular magnetic field can be established in an article in two ways: 1) by passing current directly through the article, and 2) by passing current through a _____ conductor.

A. prod
B. square
C. central
D. yoke

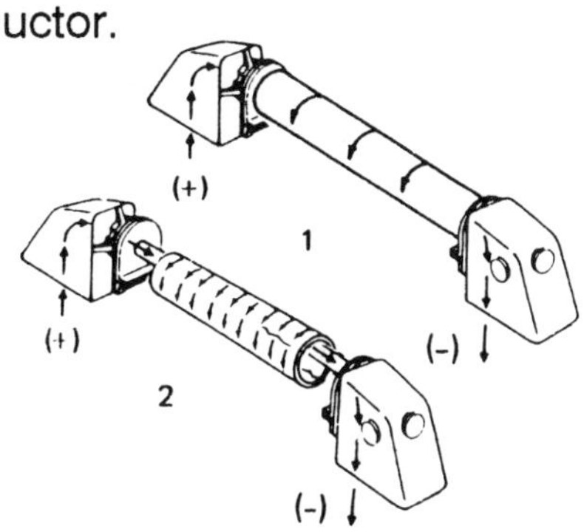

D    14. If we suspected that this gear might have cracks on either side, we would magnetize it using the _____ method.

    A.    yoke
    B.    coil
    C.    least sensitive
    D.    central conductor

C    15. A strong magnetic field is said to have greater flux density than a weak magnetic field. The higher the field strength the greater the:

    A.    permeability.
    B.    coercive force.
    C.    flux density.
    D.    retentivity.

A    16. When passing current directly through the bar (head shot), flux density is greatest at the _____ of the bar.

    A.    surface
    B.    inside
    C.    center
    D.    end

B  17. When using the central conductor, flux density is greatest at the _____ of the conductor.

   A. end
   B. surface
   C. inside
   D. center

D  18. When an article is placed inside a coil through which electric current is passing, a _____ field is set up in the article.

   A. negative reversing
   B. demagnetizing
   C. circular
   D. longitudinal

A  19. Here is a bar with a crack in it. We can establish either a circular field or a longitudinal field in this bar. Which field will cause flux leakage at the crack?

   A. Circular
   B. Longitudinal
   C. Demagnetizing
   D. Cannot be determined.

From page 2-87

_D_  20. When current flows along a copper wire, it produces a _____ in and around the wire.

   A. blue glow
   B. heat shield
   C. copper field
   D. magnetic field

_C_  21. Which type of field will cause flux leakage at this crack?

   A. Circular
   B. Remanence
   C. Longitudinal
   D. Demagnetizing

_B_  22. If we want iron particles to be attracted to this crack, we must magnetize the bar by passing current through a:

   A. prod.
   B. coil.
   C. yoke.
   D. bar.

From page 2-88

__B__ 23. When a central conductor is used to circularly magnetize an article, both ends of the article can be inspected as well as the _____ of the hollow article.

A. outside
B. inside and outside
C. inside

__A__ 24. If we want iron particles to be attracted to this crack, we must magnetize the bar by passing current through the:

A. bar (article).
B. crack.
C. coil.
D. yoke.

__D__ 25. Here is a tube with a crack on the inside surface. If we want flux leakage at that crack, we must magnetize the tube with a _____ field.

A. circular
B. coercive
C. hysteresis
D. longitudinal

From page 2-89

A   26. The flux density in a coil is _____ near the inside surface of the coil.

   A. strongest
   B. weakest
   C. non-existent

B   27. When an article is too big to fit an ordinary coil, we use a cable wrapped around the article to take the place of the ordinary coil. When electric current is passed through the cable, a _____ field is produced in the article.

   A. demagnetizing
   B. longitudinal
   C. permeable
   D. circular

B   28. The effective distance of the longitudinal magnetic field produced either by the looped cable or the ordinary coil is _____ inches on either side of the coil.

   A. 6 to 12
   B. 6 to 9
   C. 12 to 18
   D. more than 18

From page 2-90

D  29. The magnetic field set up by a pair of prods is essentially:

   A. useless.
   B. longitudinal.
   C. a leakage field.
   D. circular.

C  30. When current flows along a ferromagnetic steel bar, it produces a _____ field in and around the bar.

   A. circular demagnetizing
   B. longitudinally magnetic
   C. circular magnetic
   D. longitudinally demagnetizing

A  31. To be most effective, prods are held _____ apart.

   A. 6 to 8 inches (15 to 20 cm)
   B. 9 to 12 (23 to 30 cm)
   C. 12 to 18 (30 to 46 cm)
   D. 8 to 9 (20 to 23 cm)

C  32. A crack that runs crosswise or at least 45° to the lines of force will cause:

   A. demagnetization.
   B. discolorization of the iron particles.
   C. flux leakage.
   D. domain depletion.

___C___   33.   If we were to inspect this ring inside and out for possible seams, we would _____ magnetize it using the _____.

   A.   circularly, yoke
   B.   longitudinally, central conductor
   C.   circularly, central conductor
   D.   longitudinally, head shot

From page 2-92

## ANSWERS TO REVIEW QUESTIONS FOR CHAPTER TWO

| Question & Answer | Reference Page(s) |
|---|---|
| 1. A | 2-41 |
| 2. A | 2-1 |
| 3. C | 2-22 |
| 4. D | 2-50 |
| 5. B | 2-3 |
| 6. C | 2-81 |
| 7. A | 2-43 |
| 8. D | 2-6 |
| 9. A | 2-11 |
| 10. B | 2-27 |
| 11. D | 2-19 |
| 12. A | 2-29 |
| 13. C | 2-22 |
| 14. D | 2-51 |
| 15. C | 2-32 |
| 16. A | 2-31 |
| 17. B | 2-41 |
| 18. D | 2-55 |
| 19. A | 2-18 |
| 20. D | 2-6 |

From page 2-93

| Question & Answer | Reference Page(s) |
|---|---|
| 21. C | 2-18, 53 |
| 22. B | 2-55 |
| 23. B | 2-44 |
| 24. A | 2-6 |
| 25. D | 2-52 |
| 26. A | 2-55 |
| 27. B | 2-71 |
| 28. B | 2-69 |
| 29. D | 2-74 |
| 30. C | 2-6 |
| 31. A | 2-77 |
| 32. C | 2-18 |
| 33. C | 2-44 |

# CHAPTER 3

# MAGNETIZING CURRENTS

Electricity. It's uses and expanding horizon of applications make up a career for many scientists and engineers. We will next study, briefly, both *alternating and direct currents* and how they will be employed for magnetic particle testing.

## Alternating Current

Alternating current (AC) is the most convenient source of electrical current, since it is readily available. For this reason, and others we'll point out, it is the most widely used power source for conducting our magnetic particle tests.

In direct current (DC), the voltage and current are always in *phase* with one another. That is, current and voltage are at the same level of intensity at any one time. AC does not always behave in this fashion; in fact, AC can consist of several different phases of current. The simplest case is referred to as *single-phase,* which simply means one phase of current. The commonly used single-phase AC requires two conductors (wires) and reverses direction at the rate of 60 hertz (cycles per second) as illustrated by the AC sine curve below.

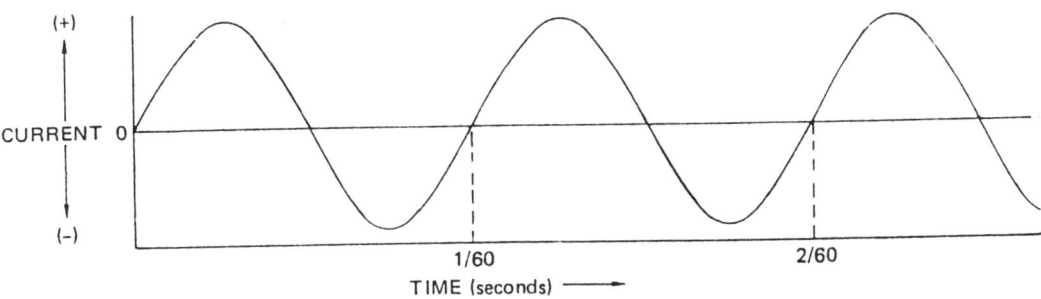

Turn to the next page.

From page 3-1                                                    3-2

Alternating current at line voltages can be stepped down efficiently and with relative ease by transformers; therefore, AC can be readily converted to the low voltages used in magnetic particle testing.

Alternating current flows on the surface of a conductor. Therefore, the magnetic field induced by alternating current is concentrated near the surface of the article being magnetized. For this reason, *AC magnetization provides the best detection of surface, or very near surface, discontinuities*. AC is not tremendously effective in detecting subsurface discontinuities. A *near surface* discontinuity is a discontinuity that lies within approximately 0.08 inches (2 mm) of the test surface.

Since AC is continuously reversing direction at the rate of 60 hertz (50 hertz in some countries), the constantly reversing magnetic field has a tendency to agitate or make the iron particles more mobile than if DC were used. AC then causes the iron particles to be more responsive to flux leakage fields than would DC.

AC also provides a ready mechanism for the demagnetization of the test article because of this reversing field. We'll look at demagnetization later in the program. Let's take a look at how we might use AC to produce either AC or DC.

Turn to the next page.

From page 3-2

3-3

## Direct Current

When single-phase AC is *rectified*, the resulting current is known as *half-wave direct current* (HWDC). This simply means that the reverse polarity or negative portion (dashed line) of the AC sine curve is eliminated.

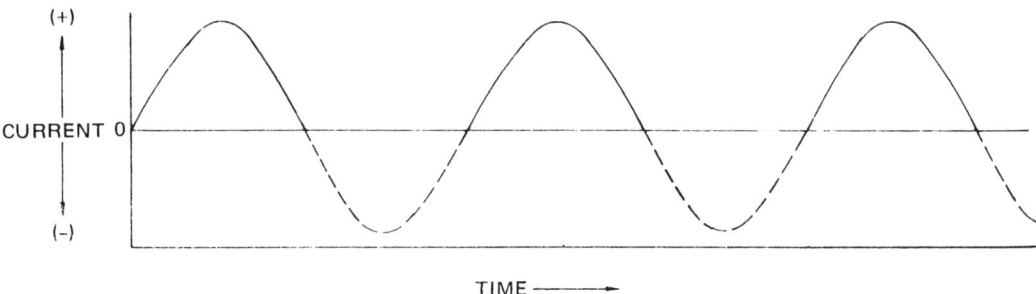

The half-wave direct current consists of individual pulses of *direct current* (DC) with time intervals in which no current is flowing. Each pulse lasts for one-half cycle, and the peak current is the same as the alternating current which is being rectified. The *average* current is considerably less than the peak current as illustrated by the dashed line below.

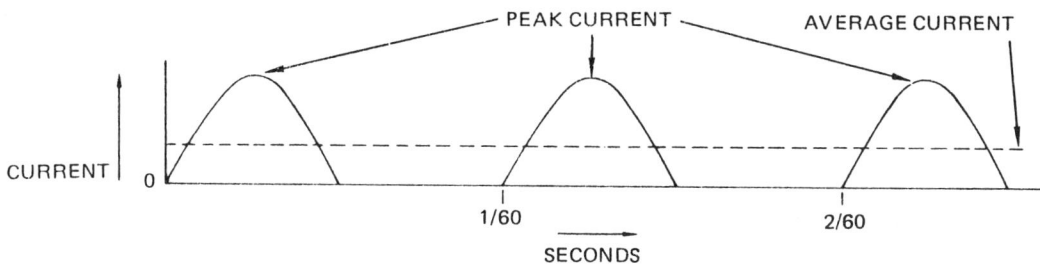

Turn to the next page.

From page 3-3                                                                 3-4

Although half-wave rectified single-phase AC is a type of direct current, it is always identified as *half-wave direct current* (HWDC). This allows differentiation between it and straight (pure) DC which is a *continuous* flow of current in one direction.

A common source of straight DC is the ordinary battery.

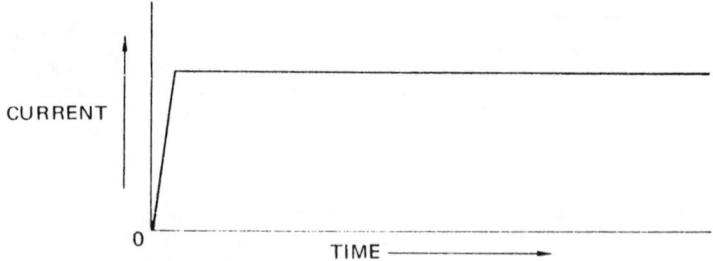

Notice that the current flow is flat in comparison to the strong pulsing current common to HWDC.

At one time, batteries were commonly used to provide DC for magnetic particle testing. However, batteries presented many problems and have now been largely replaced by other sources of DC.

Turn to the next page.

From page 3-4　　　　　　　　　　　　　　　　　　　　　　　　　　　3-5

A clue to these other sources of DC was given on the first page of our discussion of magnetizing currents. We said that AC is "the most widely used power *source* for conducting magnetic particle testing."

You have seen how single-phase AC can be rectified to give a pulsing type of DC. Now let's see what we can do with *three-phase* AC.

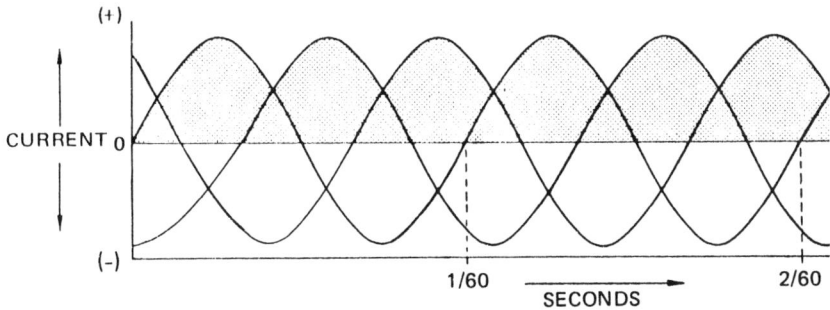

If this current is half-wave rectified, we have a pulsing DC; but the shaded portion now bears a stronger resemblance to battery DC. It is much more uniform and would have a higher average current.

**What would happen if the three-phase current were given full wave rectification? In other words, what if the negative portion of each curve were switched in such a way that it flowed in the positive direction along with the positive section of the curve?**

**It would make the positive portion of each curve higher . . Page 3-7**
**It would smooth out the ripple even more . . . . . . . . . . . . Page 3-8**

From page 3-8

Nope. You weren't paying attention. AC will establish surface and near surface magnetic fields—magnetic flux would not penetrate much beneath the surface of the specimen.

AC is optimum for detecting surface cracks where the flux leakage would be high, but *subsurface* flux density is very weak at best and would not be adequate for the detection of *subsurface* cracks.

Turn ahead to page 3-10.

From page 3-5

Sorry. Notice that the negative peaks are located vertically between the positive peaks.

If the negative current flow were reversed, the number of peaks would double—each valley between a peak would be filled with another peak as shown by the dashed curves in this illustration.

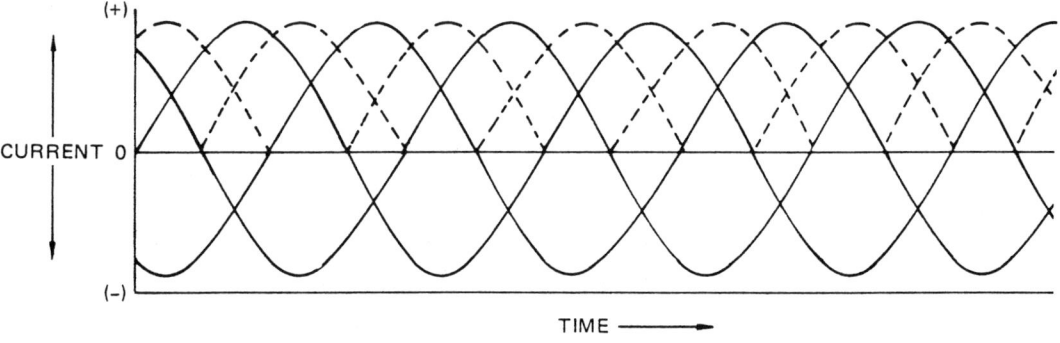

Turn to the next page.

From page 3-5  3-8

Very good. Full-wave rectification reverses the direction of the negative portion of the curve and all current flows in the same direction. We would have doubled the number of peaks on the positive curve.

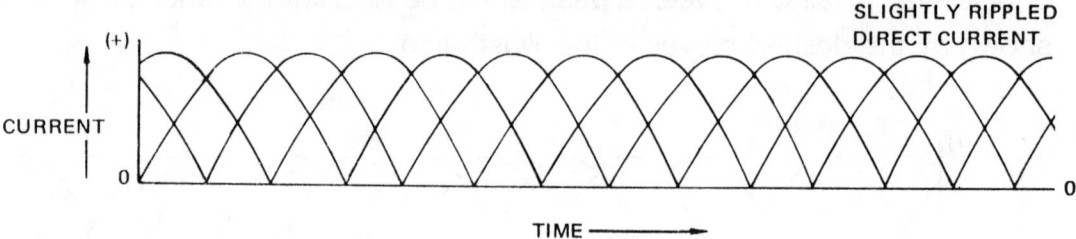

For all practical purposes, this current is the same as battery DC. It has an almost constant value with only a slight ripple.

We have stated that AC has little penetrating ability—it travels on the surface of a conductor. For that reason, the magnetic field established by AC is near surface. DC is much more penetrating; it travels within the conductor as well as on the surface. For that reason, the magnetic fields it induces are deeper.

**If you were looking for a crack beneath the surface of a specimen, which type of current would be best to use?**

**AC** . . . . . . . . . . . . . . . . . . . . . . . . . . . . . . . . . . . . **Page 3-6**
**DC** . . . . . . . . . . . . . . . . . . . . . . . . . . . . . . . . . . . **Page 3-10**

From page 3-10

The advantage in using HWDC is that a high flux density can be generated using a minimum of power. The ratio is roughly 3 to 1. For example, if an average current of 400 amps is used, the peak current will be about 1200 amps and the flux density will reflect this peak current.

Another advantage of HWDC is the strong pulsing action of the magnetic flux. This serves to agitate dry magnetic particles and makes them more responsive to flux leakage fields.

For these reasons, HWDC is often used in portable, dry particle technique equipment. It provides deeper penetration and good dry particle agitation. The combination is quite sensitive for the location of subsurface discontinuities.

**If only *surface* discontinuities are being sought, which type of current would provide the strongest flux leakage fields at the *surface* of a specimen?**

**HWDC** .................................................. **Page 3-11**
**AC** ..................................................... **Page 3-12**

From page 3-8

Absolutely. The flux density inside a specimen is much greater using DC than with AC. For example, AC can penetrate to a depth of about 0.1 inch (2.5 mm), whereas HWDC can penetrate to about 0.8 inch (20 mm) with the same magnetizing current on unhardened tool steel. Therefore, DC or HWDC should always be used for subsurface investigations.

We've now established the fact that AC can be used to detect near surface and surface discontinuities. Also, DC must be used to detect subsurface discontinuities although it will also detect surface and near surface discontinuities. We have talked about three kinds of DC; straight DC from a battery, single-phase, half-wave rectified which we have labeled "HWDC," and full-wave, three-phase rectified which we will call "FWDC."

Now let's examine the relative advantages of HWDC and FWDC. Here is our diagram of HWDC again.

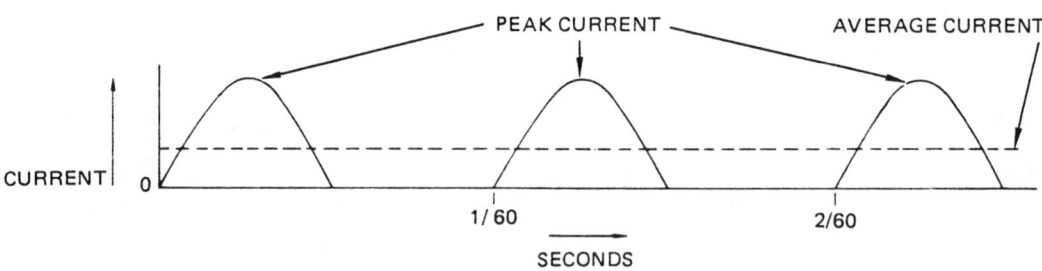

*The flux density in a specimen is determined by the peak current of the HWDC. The power requirements and heating effects are determined by the average current.* Now, can you see a possible advantage in using HWDC?

Turn back to page 3-9.

From page 3-9

You think HWDC would provide the strongest leakage field at the surface of a specimen. It is true that HWDC is used to locate both surface and subsurface discontinuities; but, since it tends to distribute itself throughout the cross section of the specimen, the flux density at the surface is not as great as with AC.

Turn to the next page.

From page 3-9                                                              3-12

Absolutely. AC, commonly at 60 Hz, provides the strongest leakage fields for surface discontinuities such as cracks or seams. Alternating current tends to flow near the surface of a conductor. Therefore, flux density is greatest at the surface when using AC.

Consider this diagram showing the magnetic field distribution when alternating current flows through a solid magnetic conductor such as a round, ferromagnetic steel bar.

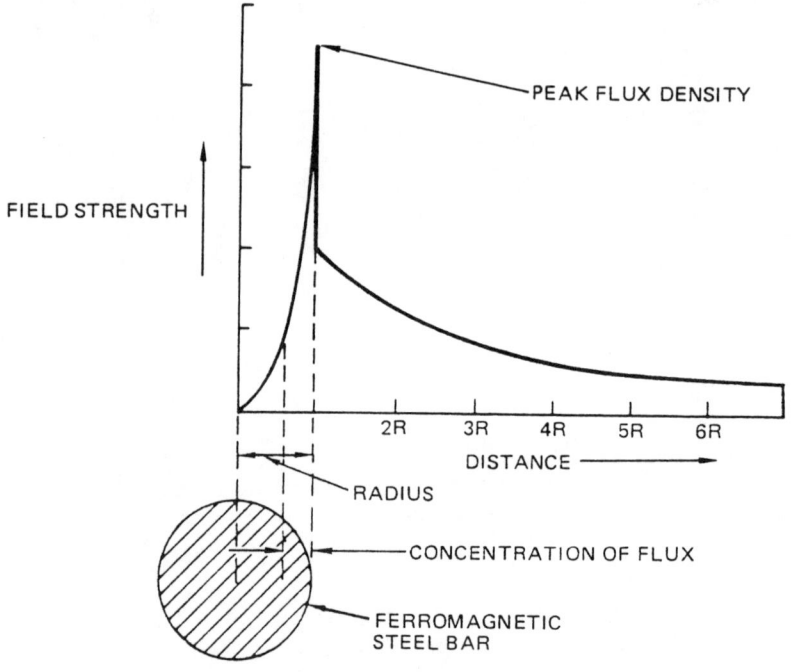

Starting at the center of the article, flux density is zero. As you can see, the greatest flux density is concentrated very near the surface of the article. It is for this reason that 60 Hz AC is widely used to detect surface and, to a limited extent, near surface discontinuities.

Turn to the next page.

From page 3-12

If you were to circularly magnetize a *hollow bar* by passing 60 Hz *AC* through its length, where do you think the current would tend to flow through the article?

**Near the inside surface (ID) of the article** ............ **Page 3-15**
**Near the outside surface (OD) of the article** .......... **Page 3-17**

From page 3-17

Right. HWDC provides the best penetration qualities for locating deep subsurface discontinuities. For a *given amperage*, HWDC will provide a stronger magnetic field than that provided by straight DC or FWDC.

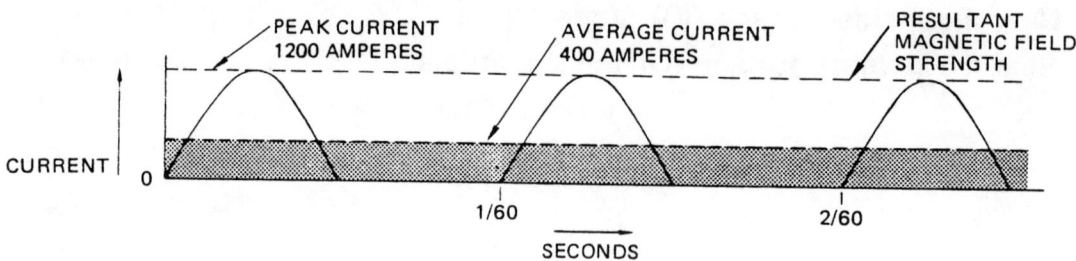

With either AC or any DC, the magnetic field varies directly with the amount of magnetizing current used. In other words, when current is increased the magnetic field strength increases. It is also true that the distribution of the electric current determines the distribution of magnetic flux. With this in mind, let us take a look at the distribution of the magnetic field in a solid, ferromagnetic steel bar through which a direct current is flowing.

Turn ahead to page 3-16.

From page 3-13                                                3-15

You think AC would tend to flow near the *inside* surface of a hollow bar? No, alternating current always tends to flow near the *outside* surface of any conductor, including a hollow bar.

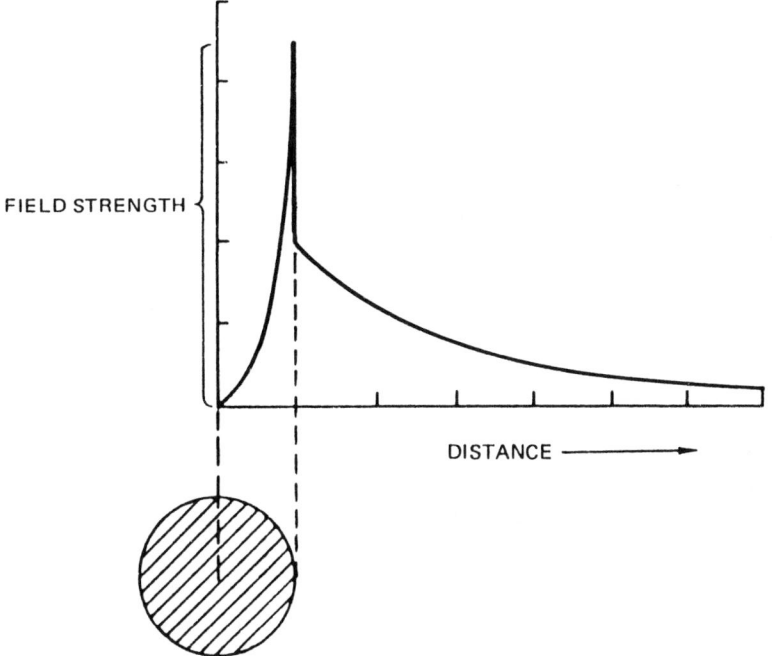

In the case of the solid piece of ferromagnetic steel, the electric current (field strength) is zero at the center. The largest portion of the current then moves rapidly toward the surface, as shown by the curved portion pointed out by the bracket in the above illustration. This same phenomenon would occur in a hollow bar also.

Turn ahead to page 3-17.

From page 3-14                                                                 3-16

DC has better penetration qualities than does AC as illustrated by the "AC" and "DC" curves below.

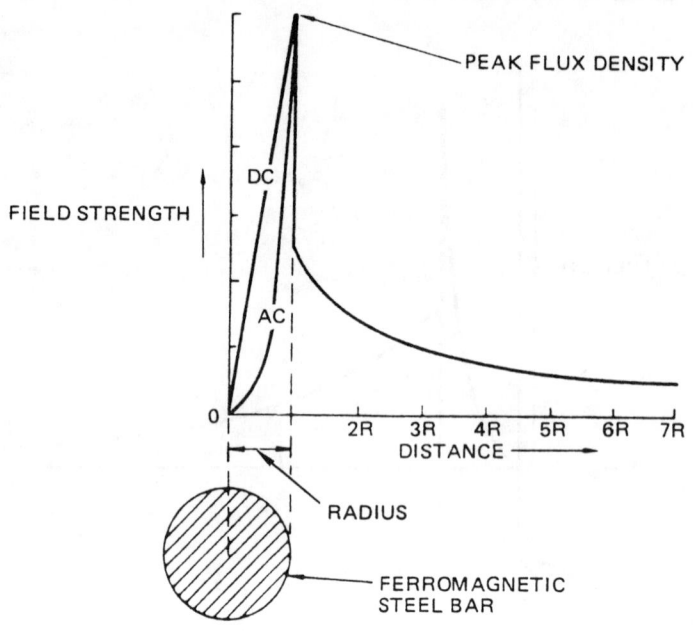

With DC, flux density increases evenly on a straight line from zero at the center of the bar to the surface. The AC curve, however, veers sharply toward the outside before the flux density increases appreciably. Since flux density also represents current density, the "AC" and "DC" curves above also represent current distribution in the bar.

**We can say that DC creates:**

**a stronger internal magnetic field than does AC** ....... **Page 3-19**
**a stronger surface magnetic field than does AC** ....... **Page 3-20**

From page 3-13                                                          3-17

Yes, 60 Hz AC will always tend to flow near the outside surface on any conductor, including a hollow bar. It is obvious that if the electric current density is concentrated in the outer layer of the conductor, the flux density will be correspondingly greater in that area. Here is a diagram showing the distribution of the magnetic field in the hollow bar through which AC is flowing.

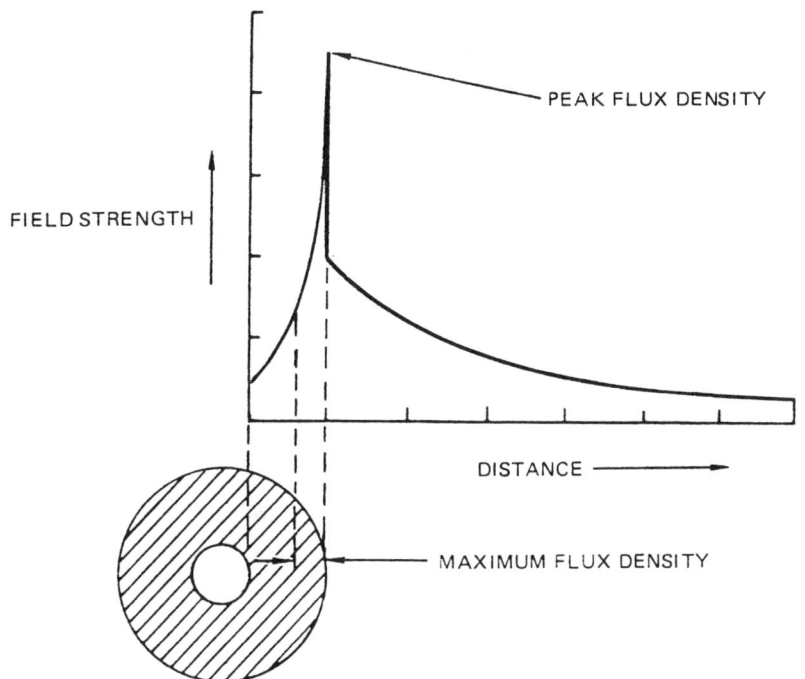

Here again, the magnetic field of the hollow bar is concentrated very near to the outer surface where the flux density is greatest. For this reason, 60 Hz AC is NOT used to locate *subsurface* discontinuities.

**Assuming the same current (amperage), which of the following types of current provides the *best* penetration qualities?**

**HWDC** .................................................. **Page 3-14**
**Straight (pure) DC** ...................................... **Page 3-18**

No. For a *given amperage*, straight DC does NOT have the best penetration qualities. Let's compare the three types of DC, remembering that the magnetizing current is equal for all three types.

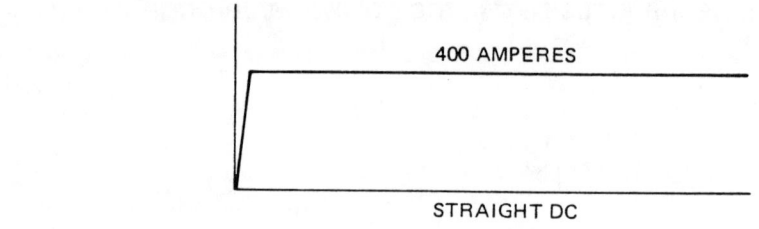
STRAIGHT DC

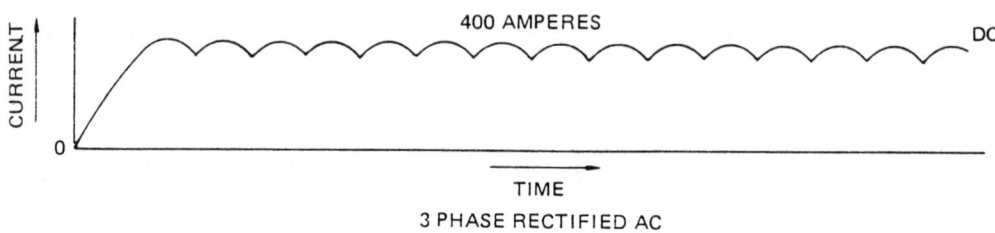
3 PHASE RECTIFIED AC

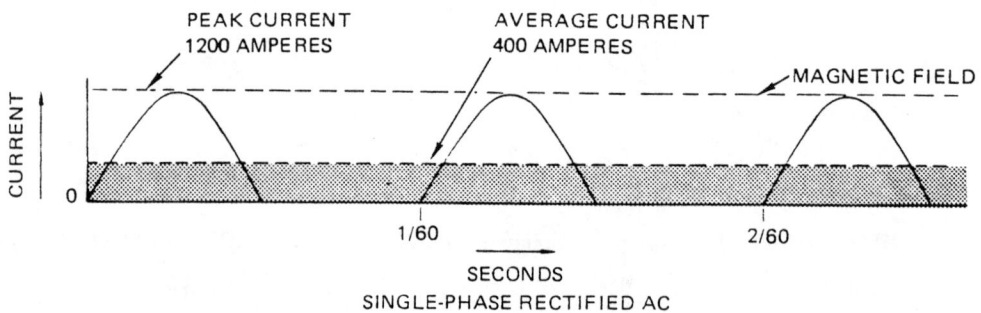
SINGLE-PHASE RECTIFIED AC

Flux density is determined by the *peak current* used. With straight DC, the peak current is 400 amps and the flux density reflects this peak. HWDC of 400 amps average value has a peak current of approximately 1200 amps. The resultant flux density will be much higher than with straight DC or FWDC (middle figure).

Turn back to page 3-14.

From page 3-16					3-19

Right. DC creates a stronger internal magnetic field than does AC. DC, and particularly HWDC, is ideal for detecting subsurface discontinuities.

Remember our previous discussion? We did say that HWDC was best for detecting subsurface discontinuities for a *given amperage*. In this discussion, we have not limited ourselves concerning amperage. Remember that AC flows along the surface of a conductor and any DC tends to flow through a conductor. That should keep you on the right track!

Obtaining DC by rectifying AC (either HWDC or FWDC) provides several advantages over battery-supplied (straight or pure) DC. These advantages are:

- Alternating 60 Hz current in either single-phase or three-phase is readily available.
- Rectifying equipment can be designed to deliver any desired current value on any required work cycle.
- The pulsating DC obtained from rectified single-phase AC (HWDC) is highly effective in obtaining indications.
- Rectified single-phase AC (HWDC) produces greater field intensities with less power used.
- Battery maintenance, fumes, hazards and disposal concerns are eliminated.

The comparative differences in sensitivity (the degree of responsiveness to magnetic particle testing) between AC and DC depend largely on the type of magnetic particles used and the technique of testing. These will be discussed in some detail later.

Turn ahead to page 3-21.

From page 3-16

No, DC does NOT create a *stronger surface* magnetic field. You may have been confused by the diagram. Let's look at it once more.

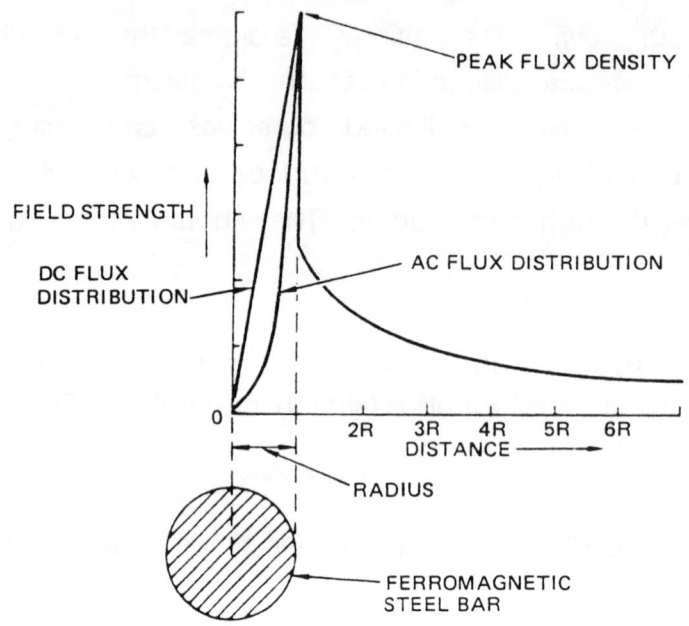

The arrow on the left is pointing to the DC magnetic field distribution in the article. The lower arrow on the right is pointing to the AC magnetic field distribution in the bar. The AC line shows a sharp outward turn which indicates that the flux density is concentrated near the outer surface of the bar. In contrast, the DC line shows an even progression to the peak flux density point. This means that the flux density is progressively increasing at a constant rate. Therefore, this indicates that DC creates a stronger internal magnetic field than does AC.

Turn back to page 3-19.

From page 3-19

## Current Requirements (Circular Magnetization)

The amount of electric current used will vary with the size and shape of the article and with the permeability of the material. For example, too much current may cause an electric arc to be struck that could burn the surface of the article. Too much current may also cause very heavy accumulations of iron particles. Too little current may not be adequate to provide sufficient flux leakage to attract iron particles.

The examiner's primary guideline for current requirements and the type of magnetizing current is the *test procedure*. The test procedure is also referred to as the specification, standard, or possibly process control sheet. The current requirements are generally given in terms of the cross-sectional dimensions of the article. A circular bar, for example, will require a certain number of amperes per inch of diameter. A rectangular bar will require a certain number of amperes per inch of largest measurement of the cross section (a diagonal measurement). Some procedures require only a largest-width measurement. Always check the applicable procedure to determine the magnetizing current requirements.

Since there are so many variables involved in determining current requirements for individual articles, we can provide only general guidelines. *For our purposes here*, let's use the following rule in determining the current needed for circular magnetization between the heads (head shot) and with the central conductor.

**USE 700 TO 1000 AMPERES PER INCH (280 TO 400 AMPERES PER CENTIMETER) OF ARTICLE DIAMETER OR DIAGONAL MEASUREMENT.**

Turn to the next page.

From page 3-21                                                3-22

One should keep in mind that as the diameter (or cross-sectional length) increases, the current requirements are reduced. But, if the part to be examined is hardened, current requirements are increased. So you can see that current requirements vary. The examiner should determine the magnetic field strength nearest the area of interest on the part by actual measurement whenever there is doubt.

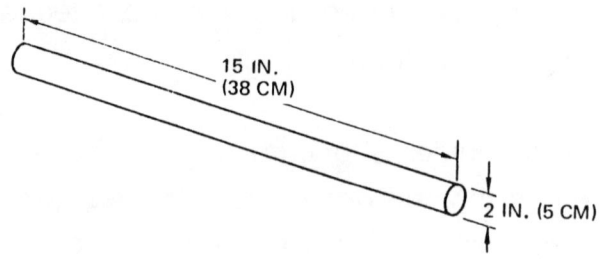

This round bar is 2 inches (5 cm) in diameter. In applying the rule, you would determine the current required (in amps) as follows:

2 inches (5 cm) x 700 amps/inch (280 amps/cm) = 1400 amps minimum
2 inches (5 cm) x 1000 amps/inch (400 amps/cm) = 2000 amps maximum

resulting in: 1400 to 2000 amperes required.

Turn to the next page.

From page 3-22                                                                 3-23

For a circular bar you simply multiply the diameter in inches by 700 and then by 1000 to get the *range* of the required amperage. Similarly, in SI units; multiply the diameter in centimeters by 280 and then by 400 to get the current range. (Note: For ease of reading, SI units may not be included in the examples.)

Now let's look at a rectangular bar.

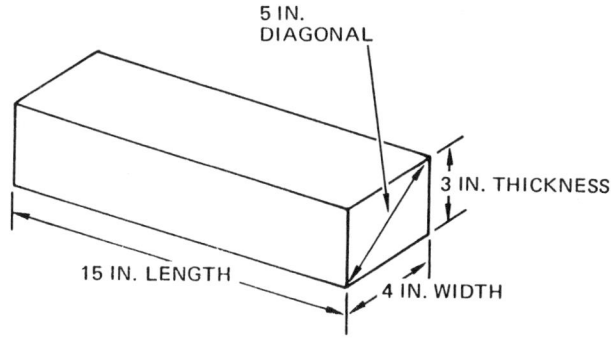

**Remembering the rule:**
    **USE 700 TO 1000 AMPERES PER INCH OF ARTICLE DIAGONAL MEASUREMENT,**
how many amperes would you use to circularly magnetize this bar? Our procedure says to use the largest measurement.

2100 to 3000 ................................. Page 3-24
3500 to 5000 ................................. Page 3-27

From page 3-23

You selected "2100 to 3000 amperes." You have mixed thickness of the bar with the diagonal measurement of the bar.

Thickness of an article is always its smallest dimension. For example, our test bar was 3 inches thick, 4 inches wide, with a *diagonal* measurement of 5 inches.

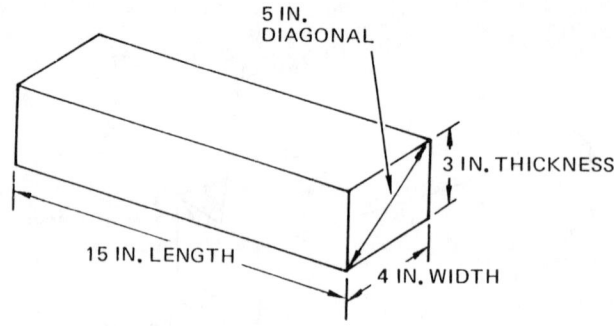

So, you can see that using the rule of 700 to 1000 amperes per inch of *diagonal* measurement would allow a range of 3500 to 5000 amperes for the above bar:

$$5 \times 700 = 3500; \quad \text{and} \quad 5 \times 1000 = 5000$$

Turn ahead to page 3-27.

From page 3-27

3-25

You chose 700 to 1000 amperes. But the bar is 3 inches wide, NOT 1 inch. *In this case*, the rule would read like this:

**USE 700 TO 1000 AMPERES PER INCH OF ARTICLE WIDTH.**

Here is the bar again.

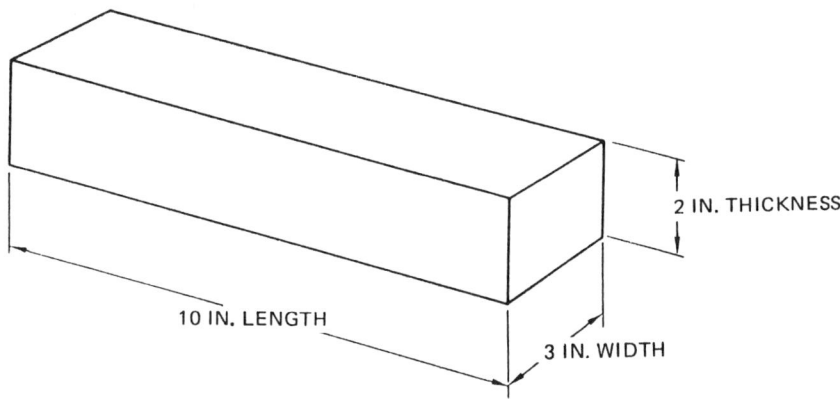

Determine amperes required as follows:

$$3 \times 700 = 2100; \quad \text{and} \quad 3 \times 1000 = 3000$$

$$\text{Amperes required} = 2100 \text{ to } 3000$$

Return to page 3-27 and select the correct answer.

From page 3-30 3-26

Your answer is correct only if the bar is 3/4 inch in diameter. Since our test bar is only 1/2 inch in diameter, we multiply like this:

$$\frac{1}{2} \times 700 = 350 \qquad \frac{1}{2} \times 1000 = 500$$

Amperes = 350 to 500

Turn ahead to page 3-32.

From page 3-23

Right. The diagonal measurement was 5 inches, so you would use an ampere range of 3500 to 5000 amperes to circularly magnetize that bar. Here is the rule again:

**USE 700 TO 1000 AMPERES PER INCH OF ARTICLE DIAMETER *OR* DIAGONAL MEASUREMENT.**

*If* the procedure requires that the *greatest width* be used to establish the current, we just multiply 700 by the number of inches of article width and 1000 by the number of inches of article width. For example, here we have a bar that is 3 inches wide.

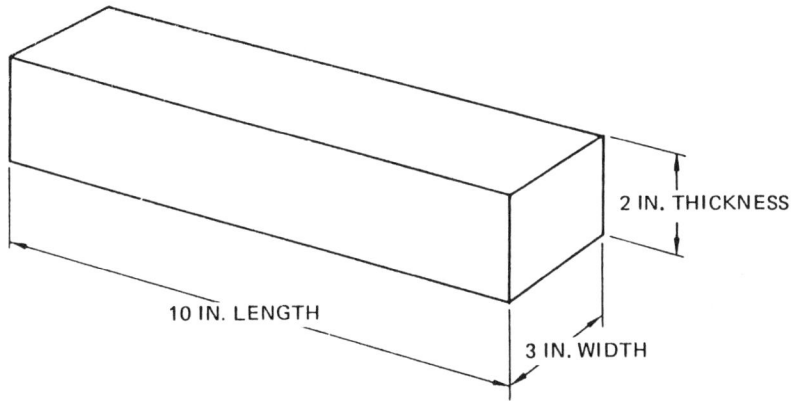

**With this procedure, what ampere range would be needed to circularly magnetize this bar?**

**700 to 1000 amperes . . . . . . . . . . . . . . . . . . . . . . . . . . . . Page 3-25**
**1400 to 2000 amperes . . . . . . . . . . . . . . . . . . . . . . . . . . . Page 3-28**
**2100 to 3000 amperes . . . . . . . . . . . . . . . . . . . . . . . . . . . Page 3-30**

From page 3-27

Ooooops! The bar is 3 inches wide, not 2 inches.

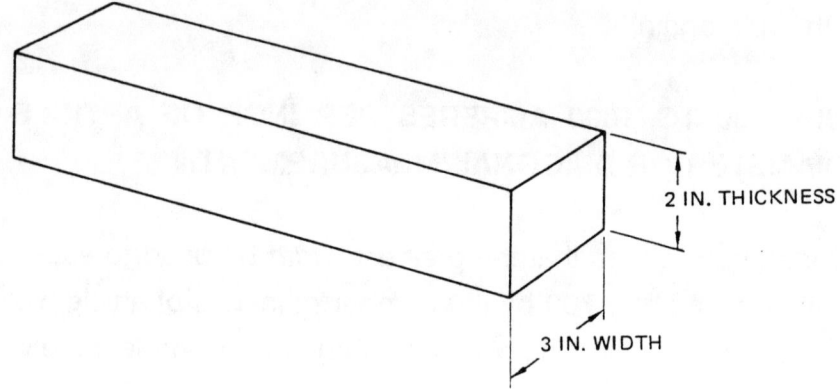

Turn back to page 3-27 and figure it out again.

You selected 700 to 1000 amperes. That would be correct if the bar was 1 inch in diameter. Since our test bar is only 1/2 inch in diameter, it requires only one-half as much current.

Turn to the next page and try again.

From page 3-27                                                                 3-30

Exactly. It would take 2100 to 3000 amperes to circularly magnetize the bar. You simply multiplied 700 x 3 = 2100 and 1000 x 3 = 3000.

Now, if the article happens to have a thickness less than 1 inch, we would use only a part of the 700 to 1000 amperes. For example, here we have a rod that is 3/4 inch in diameter.

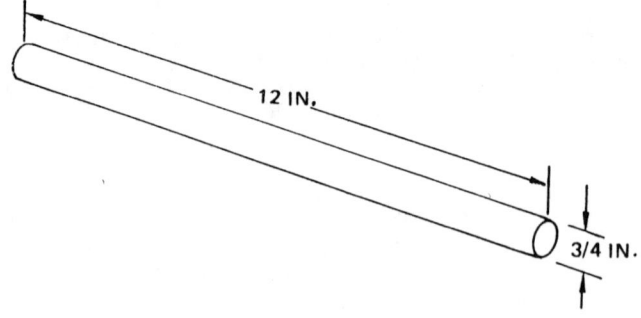

For this rod, we would use only 3/4 of 700 to 1000 amperes.

$$3/4 \times 700 = 525; \quad \text{and} \quad 3/4 \times 1000 = 750$$

From this you can see that we would use between 525 and 750 amperes to circularly magnetize the above rod.

**Now, suppose we have a bar that is only 1/2 inch in diameter. What would the amperage range be for circularly magnetizing this bar?**

525 to 750 amperes . . . . . . . . . . . . . . . . . . . . . . . . . . . . . . . . . . . . Page 3-26
700 to 1000 amperes . . . . . . . . . . . . . . . . . . . . . . . . . . . . . . . . . . Page 3-29
350 to 500 amperes . . . . . . . . . . . . . . . . . . . . . . . . . . . . . . . . . . . Page 3-32

From page 3-33

Your answer would be OK for a head shot for a 1-inch-diameter test article. Remember, we calculate the current requirements by applying the rule for head shots, BUT we must add twice the thickness of the test article to the diameter of the central conductor. Our tube has a thickness of 0.1 inch which has to be doubled then added to the conductor's diameter BEFORE we perform our calculations. You should have multiplied as follows:

$$0.5" + (2 \times 0.1") = 0.7"$$

so that the current range is:

$$700 \times 0.7 = 490 \quad 1000 \times 0.7 = 700$$

or 490 to 700 amperes.

Turn ahead to page 3-34.

From page 3-30

3-32

Very good. You would use between 350 and 500 amperes to circularly magnetize that 1/2-inch bar. Here is the rule again.

**USE 700 TO 1000 AMPERES PER INCH (280 TO 400 AMPERES PER CENTIMETER) OF ARTICLE DIAMETER**

Here is a bar already in the machine. It has an outside diameter of 12 centimeters. In the space at the bottom of this page, compute the amperage *range* required to circularly magnetize this 12 cm diameter bar.

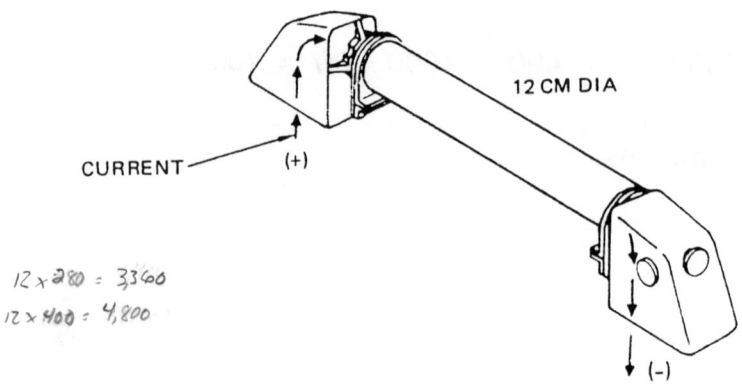

$12 \times 280 = 3,360$
$12 \times 400 = 4,800$

When you have solved the problem, turn to the next page.

From page 3-32                                                                 3-33

Your answer should be: 280 x 12 = 3360; and 400 x 12 = 4800, or to use between 3360 and 4800 amperes on the 12-cm diameter bar.

The rule of using 700 to 1000 amperes per inch of article diameter also applies to circular magnetization with a central conductor. In cases where the test article just fits over the central conductor, our rule applies directly.

When testing hollow articles that "hang" loosely on a central conductor, we calculate the current requirements by applying the rule above to the diameter of the central conductor after adding twice the test article's thickness: D + 2T. (Again we remind you to check the actual test procedure for the current requirements.)

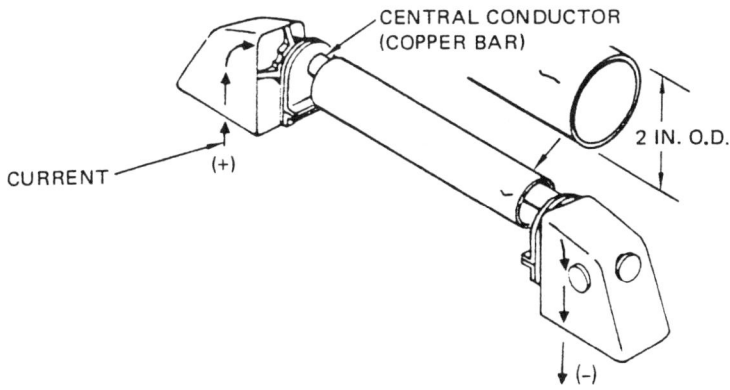

The above central conductor has a diameter of 0.5 inch. The tube suspended on it has a thickness (T) of 0.1 inch. **What would be the correct ampere range for this tube?**

350 to 500 amperes . . . . . . . . . . . . . . . . . . . . . . . . . . . . . Page 3-31
490 to 700 amperes . . . . . . . . . . . . . . . . . . . . . . . . . . . . . Page 3-34

Right. The correct amperage for that 0.1 inch thick hollow tube on a 0.5 inch central conductor would be 490 to 700 amperes. As we said, the rule **USE 700 TO 1000 AMPERES PER INCH (240 TO 400 AMPERES PER CENTIMETER) OF ARTICLE DIAMETER** is only a guide. If exceptionally heavy accumulations of particles appear, especially at abrupt changes of article thickness, the suggested amperage should be reduced. If, on the other hand, the accumulation of particles looks sparse, the amperage may need to be increased.

When using a central conductor to magnetize an article, always use the *largest conductor* that is practical for the situation, since the diameter of the conductor does have an affect on the magnetizing field produced. The effective field of the central conductor is approximately four times the diameter of the conductor, as illustrated below.

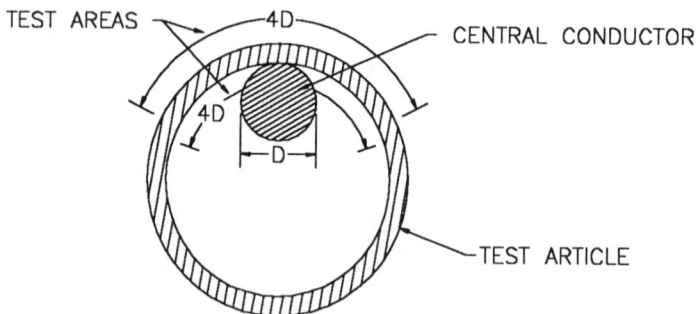

For cylindrical articles having a large diameter with respect to the central conductor, it is necessary to reposition the article on the conductor and re-inspect allowing for approximately 10 percent overlap of the magnetic field. In our previous example, this means that 4 times a diameter of 0.5 inch, or a circumferential length of 2 inches, is the maximum coverage for a single test. If the test article has a diameter of 2 inches (a circumference of over 6 inches), at least 4 rotations of the test article are required to complete the examination.

Turn to the next page.

From page 3-34

3-35

## Current Requirements With Prods

When using prods, the electric current used will vary with the following:

- the thickness of the material.

- the distance between prods.

- current limitations imposed by the heating of the article at the prod contacts; limitations of the power source; and flux leakage caused by the field which is around the prods and external to the article.

Most of the time we can control the prod spacing; and when we cannot, it is often because some test access restriction forces us to place the prods closer together.

> **Note**: Prod spacing of less than 2 inches is NOT to be used. Additionally, laboratory experiments have shown that prod spacing should NOT exceed 8 inches. When the prod spacing exceeds 8 inches, the amount of current required is such that the iron filings tend to migrate to the prod contact points. Also, the leakage fields resulting from discontinuities between the prods is significantly diminished.

Turn to the next page.

From page 3-35

The thickness of the material and prod spacing are the controlling factors in selecting the current to be used up to the limitations just mentioned. The table below gives the amperage ranges for various prod spacings and section thicknesses.

| PROD SPACING INCHES (CM) | SECTION THICKNESS, INCHES (CM) | |
| --- | --- | --- |
| | UNDER 3/4 (2) | 3/4 (2) AND OVER |
| 2 to 4 (5 to 10) | 200 to 400 amps | 230 to 460 amps |
| Over 4 (10) to less than 6 (15) | 400 to 600 amps | 460 to 690 amps |
| 6 to 8 (15 to 20) | 600 to 800 amps | 690 to 920 amps |

Turn to the next page.

## Current Requirements (Longitudinal Magnetization)

The amount of current needed to *longitudinally magnetize* a test article positioned through and toward the side of a coil is calculated using the following equation:

$$\frac{K}{L/D} = Ampere\text{-}turns$$

where:

L = length of the article
D = diameter of the article

and:

K = a constant *established by the procedure* applicable to the particular test. This constant remains the same for all tests conducted under that procedure. When the test article is positioned to the side of the coil, we use K = 45,000.

L/D is the length-to-diameter or thickness ratio of the article. It may be expressed as: length over diameter equals the L/D ratio.

$$\frac{Length}{Diameter} = L/D$$

Turn to the next page.

In other words, the length of an article divided by the diameter will give the L/D ratio. For example:

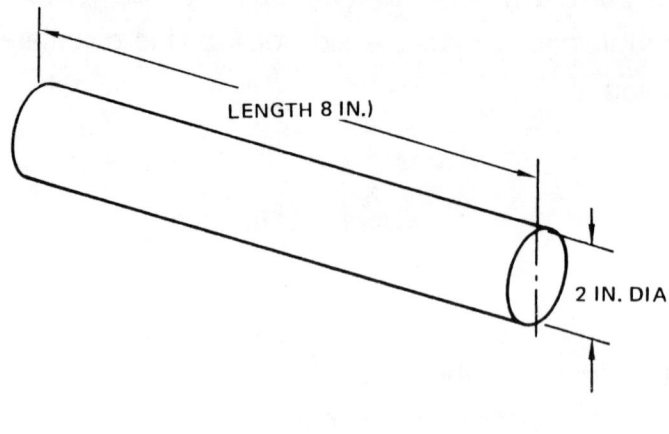

$$\frac{Length}{Diameter} = \frac{8}{2} = 4$$

The length of the article, 8 inches, divided by the diameter, 2 inches, equals 4, which is the L/D ratio.

Turn to the next page.

From page 3-38

Here is the equation again:

$$\frac{45,000}{L/D} = Ampere\text{-}turns$$

The L/D ratio of an article is determined by dividing the length of an article by the diameter of the article. In the space at the bottom of this page, calculate the L/D ratios for the following articles. Enter the figures in the L/D ratio column.

| Article Length | Article Diameter | L/D Ratio |
|---|---|---|
| 9 in. | 3 in. | 3 |
| 10 cm | 5 cm | 2 |
| 14 in. | 1 in. | 14 |
| 40 cm | 8 cm | 5 |
| 18 in. | 1½ in. | 12 |

When you have computed the above L/D ratios, turn to the next page and check your answers.

From page 3-39

Here are the correct answers.

| Article Length | Article Diameter | L/D Ratio |
|---|---|---|
| 9 in. | 3 in. | 3 |
| 10 cm | 5 cm | 2 |
| 14 in. | 1 in. | 14 |
| 40 cm | 8 cm | 5 |
| 18 in. | 1½ in. | 12 |

Notice that the above computations are based on articles that have a *length no greater than 18 inches* (46 cm). As mentioned earlier, the effective length of a longitudinal magnetic field is 6 to 9 inches (15 to 23 cm) on either side of a coil. An article with a length greater than 18 inches will require two or more coil shots.

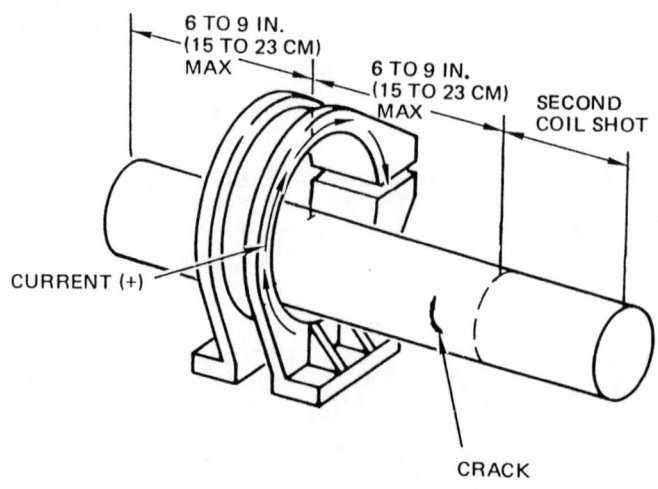

Turn to the next page.

From page 3-40

Now that we have reduced the L/D ratio to a number, we can continue with the equation to determine the current required for longitudinal (coil) magnetization. Here is the equation again.

$$\frac{45{,}000}{L/D} = Ampere\text{-}turns$$

Let's work with an article 10 inches long with a diameter (or thickness) of 5 inches.

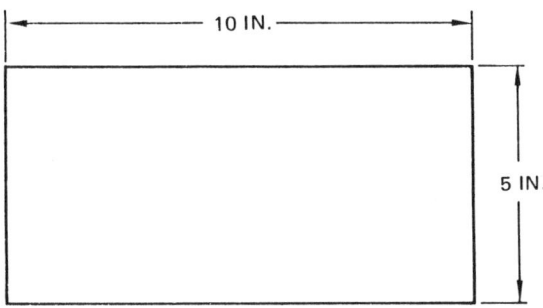

The L/D ratio is 2.  $\dfrac{Length}{Diameter} = \dfrac{10}{5} = 2$

The L/D ratio of 2 is used in the equation in place of the L/D symbols.

$$\frac{45{,}000}{2} = Ampere\text{-}turns$$

To determine the ampere-turns, we simply divide 45,000 by the L/D ratio of 2. Determine the ampere-turns in our equation now.

$$\frac{45{,}000}{2} = 22{,}500$$

Turn to the next page and check your answer.

From page 3-41

Your answer should be: $\dfrac{45{,}000}{2} = 22{,}500$ *Ampere-turns*

Very good. Let's try another example.

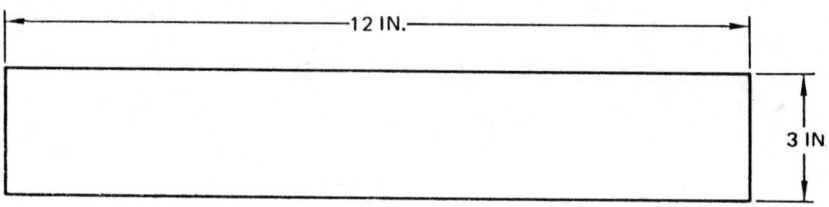

Using the equation, determine the ampere-turns for the above article.

$$\dfrac{45{,}000}{L/D} = Ampere\text{-}turns$$

In the space below, compute the answer to the above problem.

$$\dfrac{45{,}000}{4} = 11{,}250$$

When you have completed the problem, turn to the next page and check your answer.

From page 3-42

Your answer should be computed as follows:

$$\frac{L}{D} = \frac{12}{3} = 4$$

so that:
$$\frac{45,000}{4} = 11,250 \text{ Ampere-turns}$$

If your answer is not correct, return to the previous page and recheck your figures.

All of the previously computed values gave us the number of *ampere-turns* required. To determine the amperage (current) required, we divide the number of ampere-turns by the number of turns (or loops) in the coil. Most coils have from 3 to 5 turns in them. This information is marked on some coils; if not, it should be obtained from the manufacturer of the coil and recorded for use. Let's assume that our coil has 5 turns (loops). Using the figure from the example above, let's compute the magnetizing current.

$$\frac{45,000}{4} = 11,250 \text{ Ampere-turns}$$

If the coil has 5 turns, then:

$$\text{Magnetizing Current} = \frac{11,250}{5} = 2,250 \text{ Amperes}$$

What would be the magnetizing current if our coil had only 3 turns? Compute the answer below.

*3750*

Turn to the next page to check your answer.

From page 3-43

Your answer should be as follows:

$$\text{Magnetizing Current} = \frac{11{,}250}{3} = 3{,}750 \text{ Amperes}$$

Compute the magnetizing current needed for the article illustrated below using a five-turn coil.

$$\text{Formula:} \quad \frac{45{,}000}{L/D} = \text{Ampere-turns}$$

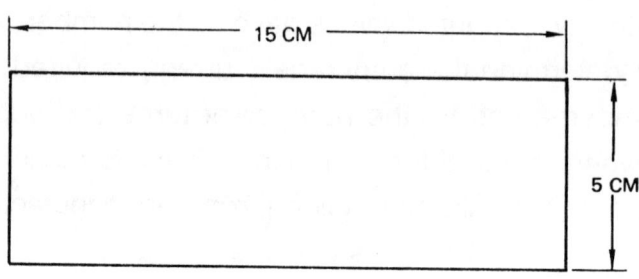

$$\frac{45{,}000}{3} = \frac{15{,}000}{5} = 3{,}000$$

Turn to the next page to check your answer.

From page 3-44

Your answer should be computed as follows:

$$\frac{L}{D} = \frac{15}{5} = 3 \quad \text{and;} \quad \frac{45,000}{3} = 15,000 \text{ Ampere-turns}$$

Using a five-turn coil: $\quad Magnetizing Current = \frac{15,000}{5} = 3,000 \text{ Amperes}$

Determine the magnetizing current for an article that is 16 inches long with a diameter of 2 inches. Assume that you are using a 5-turn coil.

$$\frac{45000}{8} = 5,625$$

$$\frac{5625}{5} = 1125$$

Turn to the next page to check your answer.

Your answer should be calculated as follows:

$$\frac{L}{D} = \frac{16}{2} = 8 \quad \text{and,} \quad \frac{45{,}000}{8} = 5{,}625 \text{ Ampere-turns}$$

Using a 5-turn coil we obtain:

$$\text{Magnetizing Current} = \frac{5{,}625}{5} = 1{,}125 \text{ Amperes}$$

If your answer was not correct, return to the previous page and recheck your figures. The use of the formula we've been working with is based upon several assumptions or limitations as follows:

- An article greater than 18-inches (46 cm) long requires more than one coil shot.

- The cross section of the article is not greater than one-tenth of the area of the coil opening.

- The article has an L/D ratio of between 2 and 15. We can in fact still use the formula for articles that exceed the upper limit by substituting the maximum of 15 for the L/D ratio.

- The article is placed against the inside wall of the coil and NOT positioned in the center of the coil.

The last point above is very important because the magnetic field strength is greatest at the inside wall of the coil. There is essentially no flux at the center of the coil.

Turn to the next page.

The lines of flux around the coil are concentrated close to the inside of the coil. As a result, the flux density is greatest near the inside wall of the coil.

In the center of the coil there is little flux, so an article which is to be magnetized is always placed such that it is in contact with or near the inside wall of the coil.

> **Note:** When an article is placed in a coil and the current turned on, the article will be attracted rather sharply to the wall. This can be disconcerting if you are not prepared for it. If the article is not attracted, then it is *not* ferromagnetic material and cannot be magnetic-particle tested.

Turn to the next page.

Now that you fully understand how to use the previous formula to calculate current requirements, we must remind you that the previous formula was for test articles *significantly* smaller than the inside area of the coil.

If we have an article to test that is only *slightly* smaller than the inside area of the coil, we should apply the following equation:

$$\text{Ampere-turns} = \frac{35{,}000}{(L/D) + 2}$$

where the L/D ratio is computed in exactly the same manner as before.

You easily realize that the only real difference is the constant on the top of the equation, and the fact that we add 2 to the L/D ratio before dividing. Not to worry, we *won't* run through any more examples!

It is, however, important to point out that other formulas are used to calculate longitudinal current requirements depending on the test article's shape and size and the cross-sectional area of the coil. The specific formula to use will be given in the test procedure.

Remember, not all test articles will allow the use of such specific formulas to calculate current requirements. This is due often to unusual shape or abrupt thickness changes. It is often standard practice to determine the magnetic field strength in the area of interest on the part by actual observation and measurement.

Relax for a bit, then turn to the next page for a chapter review.

## CHAPTER REVIEW

_B_ 1. Here is a round, ferromagnetic steel bar. We have been asked to test it for any possible cracks. First we will magnetize it between the heads (head shot). What current range, in amps, will be required?

*(Diagram: a cylindrical bar, 18 IN. long and 2 IN. in diameter)*

- A. 700 to 1000
- B. 1400 to 2000
- C. 2100 to 3000
- D. Can not be determined

_A_ 2. When amperage is the same, HWDC provides the strongest subsurface magnetic field and is the best type of current for locating _____ discontinuities.

- A. subsurface
- B. near surface
- C. surface
- D. welding

___C___   3.   Let's magnetize this bar in a 5-turn coil. How much current, in amps, will be required?

A.   9000
B.   5000
C.   1000
D.   700

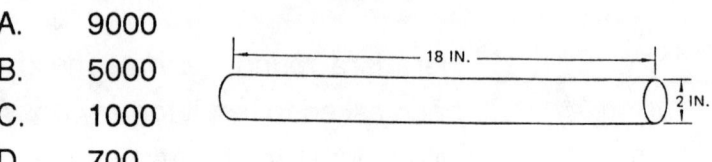

___D___   4.   Here is a rod that has been machined from bar stock. The original stock had a seam in it and we have been asked to determine if the seam was removed by the machining process. How would you magnetize the bar and how much current, in amps, would you use?

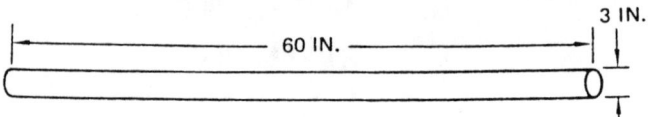

A.   With a head shot with 700 to 1000
B.   With a head shot with 1400 to 2000
C.   With a coil with 1400 to 2000
D.   With a head shot with 2100 to 3000

___A___   5.   What is the current required for each shot using a 5-turn coil on this axle which is small when compared to the opening of the coil?

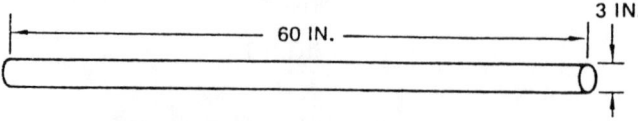

A.   (45000/15)/5 = 600 amps
B.   (35000/20)/5 = 350 amps
C.   (45000/20)/5 = 450 amps
D.   (35000/17)/5 = 412 amps

From page 3-50

A 6. Because AC tends to flow near the surface of an article, _____ is greater near the surface of the article.

- A. flux density
- B. DC
- C. permeability
- D. rectification

D 7. Here is a 12-inch section of a 4-inch pipe. We are asked to determine if there are any seams on the inside of the pipe. How would you magnetize the pipe and how much current would you use?

- A. Head shot, 2800-4000 amps
- B. Prods, 3400 amps
- C. Yoke, 1000 amps
- D. Central conductor, 2800-4000 amp

B 8. We are asked to determine if there are any circumferential cracks on the inside surface of a pipe. How would you magnetize it?

- A. Head shot
- B. Coil shot
- C. Yoke
- D. Prods

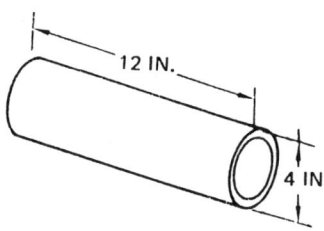

From page 3-51                                                      3-52

__D__  9.  Common alternating current (AC) reverses polarity at the rate of _____ hertz.

  A.  50 to 60
  B.  500
  C.  600
  D.  50 or 60

__C__  10.  The alternating nature of AC is illustrated by this AC _____ curve.

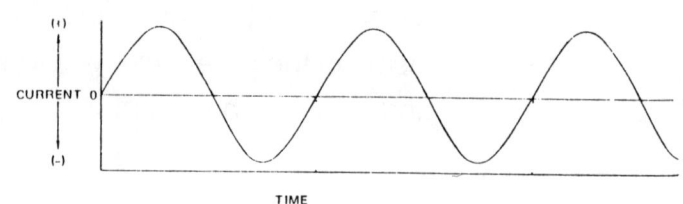

  A.  cosine
  B.  tangent
  C.  sine
  D.  cosecant

__B__   11.  If we were asked to inspect this rod for axial cracks, we would "shoot" it with a coil. What is the minimum number of shots that would be required? What is the maximum number of shots required?

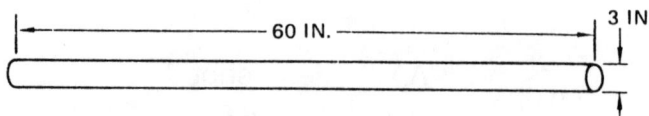

  A.  5, 4
  B.  4, 5
  C.  3, 4
  D.  1, 2

C   12.   When singe-phase AC is rectified, the negative polarity portion of the curve is:

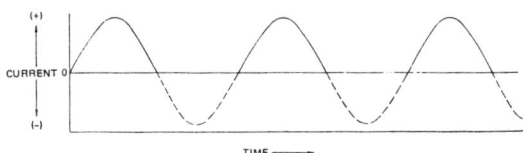

   A.   added on top.
   B.   doubled.
   C.   eliminated (removed).
   D.   called three-phase.

A   13.   HWDC consists of individual pulses of _____ current.

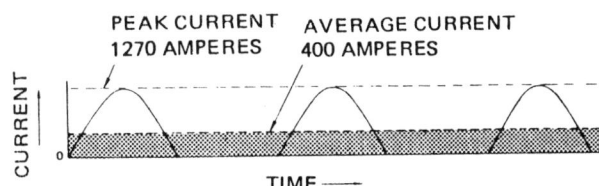

   A.   direct
   B.   alternating
   C.   battery
   D.   longitudinal

C   14.   When three-phase, alternating current is rectified, for practical purposes it can be considered to be straight:

   A.   alternating current.
   B.   single-phase alternating current.
   C.   direct current.
   D.   three-phase direct current.

From page 3-53

B     15.  Batteries are no longer in common use as a source of _____ for magnetic particle inspection.

  A. alternating current
  B. direct current
  C. single-phase direct current
  D. three-phase direct current

D     16.  Because AC flux density is greatest near the surface of an article, it is the most effective for detecting _____ discontinuities.

  A. surface
  B. subsurface
  C. near surface
  D. A and C above

C     17.  Given the same magnetizing amperages for the three types of direct current (straight DC, HWDC, and FWDC) _____ provides the greatest penetration qualities.

  A. AC
  B. DC
  C. HWDC
  D. FWDC

From page 3-54                                                      3-55

__D__  18.  Since direct current is distributed more evenly over the cross section of an article being magnetized, DC provides a stronger subsurface _____ field than does AC.

   A. amperage
   B. three-phase
   C. rectified
   D. magnetic

__A__  19.  The continuous cycle pulsing of HWDC tends to provide a vibratory movement or mobility to dry magnetic particles which aids in their attraction to weak _____ fields.

   A. leakage
   B. rectified
   C. three-phase
   D. amperage

__B__  20.  When the negative cycle of the AC curve is eliminated, the resulting current is often called half-wave _____ current.

   A. alternating
   B. direct

From page 3-55

_C_  21. In contrast to DC, 60 Hz alternating current flows at and near the _____ of an article.

   A. bottom
   B. center
   C. surface
   D. edges

_A_  22. When DC is used for magnetic particle testing, the peak current determines the resultant _____ field.

   A. magnetic
   B. rectified
   C. amperage
   D. three-phase

_B_  23. Because of the continuous cyclic reversing of polarity, 60 Hz AC also causes a _____ movement of magnetic particles which aids in their attraction to leakage fields.

   A. planar
   B. vibratory
   C. triangular
   D. polar

A    24.    DC has better penetrating qualities than _____ current.

A. alternating
B. HWDC
C. FWDC
D. battery

# ANSWERS TO REVIEW QUESTIONS FOR CHAPTER 3

| Question & Answer | | Reference Page(s) |
|---|---|---|
| 1. | B | 3-21 |
| 2. | A | 3-9 |
| 3. | C | 3-37, 43 |
| 4. | D | 3-23 |
| 5. | A | 3-46 |
| | | |
| 6. | A | 3-6 |
| 7. | D | 3-21 |
| 8. | B | 3-36 |
| 9. | D | 3-2 |
| 10. | C | 3-1 |
| | | |
| 11. | B | 3-34 |
| 12. | C | 3-3 |
| 13. | A | 3-3 |
| 14. | C | 3-5 |
| 15. | B | 3-4 |
| | | |
| 16. | D | 3-10 |
| 17. | C | 3-10 |
| 18. | D | 3-9 |
| 19. | A | 3-9 |
| 20. | B | 3-3 |

From page 3-58

| Question & Answer | | Reference Page(s) |
|---|---|---|
| 21. | C | 3-6 |
| 22. | A | 3-9 |
| 23. | B | 3-9 |
| 24. | A | 3-2 |

# CHAPTER 4

# MATERIALS AND SENSITIVITY

## Particles

Particles used in magnetic particle testing are made of carefully selected ferromagnetic materials of proper size, shape, and magnetic permeability. These particles retain practically no residual magnetism. The particles are much smaller than iron filings. In fact, when the particles are dry, they are in a flour-like powder form. Particle size is a factor in the ability of the magnetic particle test to indicate surface or subsurface discontinuities. In general, the particle size should be no greater than the surface width of the smallest rejectable discontinuity. Particle size ranges from 7.87 $\mu$ inches (0.2 $\mu$m) up to 0.016 inch (0.4 mm). The symbol $\mu$ represents micro, meaning one millionth.

Magnetic particles are provided in several colors such as black, blue, red, grey, and yellow-green fluorescent. These colors and particle sizes vary with manufacturer and application. The particles are of different mediums in accordance with the way they are used, either *wet* or *dry*.

Turn to the next page.

With the wet bath medium, the particles are suspended in a liquid vehicle. The liquid may be either water or oil. The bath is stirred or agitated to keep the particles evenly distributed in the liquid. The bath is also pumped through a hose so that it may be directed over the article to be magnetized as illustrated above. Wet bath mediums are also supplied in aerosol cans.

Magnetic particles for the wet bath medium are provided in black, grey, red, and a yellow-green fluorescent. The black and red particles provide a color contrast against the background of the article to be magnetized. The particles in the bath will be attracted to flux leakage but, when no flux leakage exists, the particles will flow off the article with the bath. An accumulation of the particles at areas of flux leakage provides an indication of a discontinuity.

Turn to the next page.

From page 4-2                                                              4-3

Fluorescent particles for the wet bath medium are used with an almost invisible light called *black light*. These particles are visible only under black light with wavelengths between visible and ultraviolet light (3200 to 4000 angstroms). Black light causes many materials, such as the *fluorescent particles*, to glow in the dark. This fluorescence is normally a brilliant yellow-green glow that "catches" the eye of the examiner.

When viewed under black light, fluorescent particles that have accumulated at areas of flux leakage will glow with great brilliance and provide an indication of a discontinuity. The main advantage of fluorescent particles is their increased visibility under black light.

Turn to the next page.

From page 4-3

4-4

Magnetic particles of high permeability are required to assure that even weak leakage fields will attract and hold the particles. In general, low retentivity particles are required so they will lose their magnetism. In this way, they are easily removed from the article if they are not held by a leakage field. Particles with these qualities would have a hysteresis loop like this.

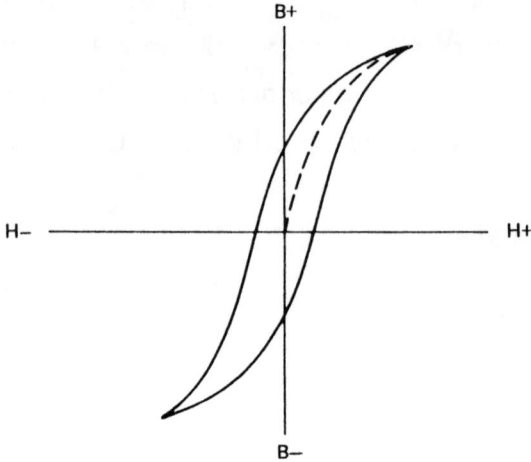

**From the hysteresis loop, you can tell that magnetic particles used in magnetic particle testing would:**

require a very high coercive force . . . . . . . . . . . . . . . . . . . Page 4-6
require high retentivity . . . . . . . . . . . . . . . . . . . . . . . . . . Page 4-8
require a very low coercive force . . . . . . . . . . . . . . . . . . Page 4-10

From page 4-10

The number of magnetic particles in the bath is called its *strength* or *concentration*. If the bath strength is not at the proper level, testing cannot be reliable. If too few particles are in the bath, no indications will be obtained. If there are too many particles in the bath, indications will be masked. If the concentration of particles in suspension in the bath varies, the strength of the indications will also vary thereby making it very difficult to accurately interpret the indications. Since the proper concentration of particles in the bath is so important, let's briefly discuss the procedures involved in bath preparation.

Cleanliness of the equipment and the bath is vital for reliable testing. Any foreign materials, such as lint, grinding dust, shop dust, and dirt or oil from parts not properly cleaned, may interfere with the proper distribution of the bath over the article. They may also change the character or definition of the deposit of the particles at an indication, or they may cause nonrelevant indications. Special attention should therefore be given to cleanliness, as well as accuracy in mixing. If at any time the bath becomes contaminated, it should be changed. The following steps should be performed.

- Before mixing a new bath, the equipment must be cleaned thoroughly. Remove and clean all the pipes and hoses in the unit and the tank, pump, and strainer if so equipped. You may choose to repeatedly flush the system with the bath to be used (oil or water, depending on system).

- After the tank and hoses have been cleaned, close all drain valves and fill the tank with the proper bath as required by the procedure.

Turn ahead to page 4-7.

From page 4-4    4-6

You feel that magnetic particles would require a very high coercive force. You must have forgotten the definition of coercive force.

**COERCIVE FORCE** is defined as:

**THE REVERSE MAGNETIZING FORCE REQUIRED TO REDUCE THE RESIDUAL MAGNETISM TO ZERO.**

On the hysteresis loop, the coercive force is shown at two points.

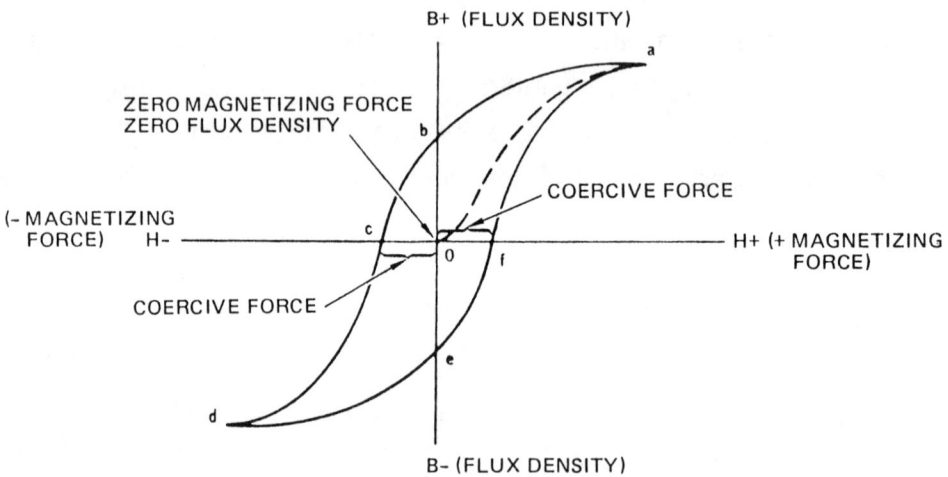

The greater the coercive force, the "harder" the material and the greater will be the residual magnetism in the material. The wider the hysteresis loop, the harder the material and lower the permeability. Since magnetic particles used in magnetic particle inspection must have high permeability, a much thinner hysteresis loop is indicated. What happens to the coercive force with a very thin hysteresis loop?

Turn back to page 4-4 and select the correct answer.

From page 4-5

- Measure the weight of the particles required by the procedure into a clean container. The amount (weight) used may vary depending on the particle manufacturer, but typical quantities are:

    - 5 pounds (2.27 kg) of nonfluorescent particles per 100 gallons (378 liters) of oil.

    - 1/4 pound (113 gm) of fluorescent particles per 100 gallons (378 liters) of oil.

- Turn on the tank's circulating pump motor. Slowly pour the particles into the tank.

- Let the pump run to mix the particles in the bath for at least 30 minutes to assure complete particle distribution.

- Flow the bath through a hand-held hose and/or nozzle for 3 to 5 minutes. This assures that the bath delivery system will contain the newly mixed bath.

Turn ahead to page 4-9.

From page 4-4

You think magnetic particles require high retentivity?  NO!  High retentivity means that the material *retains* a strong residual magnetic field.  Magnetic particles could NOT be removed from an article if they had high retentivity.

A thin hysteresis loop (left) indicates a material of low retentivity and a wide hysteresis loop (right) represents material with high retentivity.

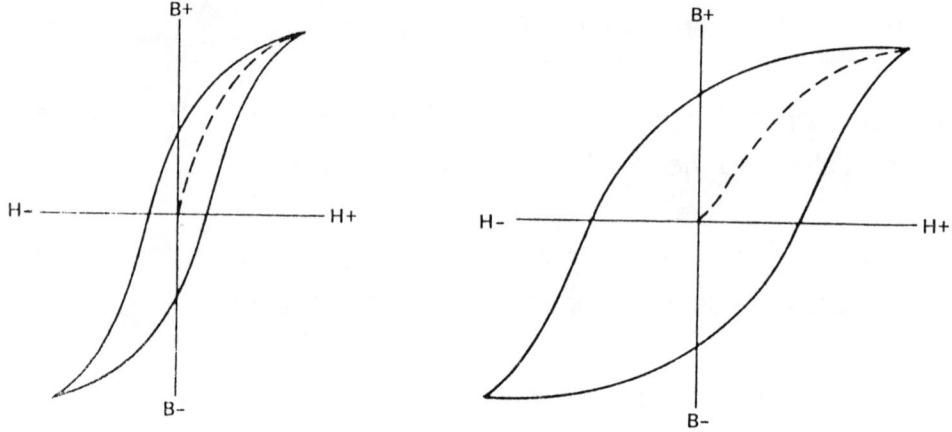

Turn back to page 4-4 and select one of the other answers.

From page 4-7                                                                 4-9

- Fill a centrifuge tube to the 100 cubic centimeter (100 ml) line as illustrated. (Other means of filling the centrifuge tube with bath may be specified in the procedure.)

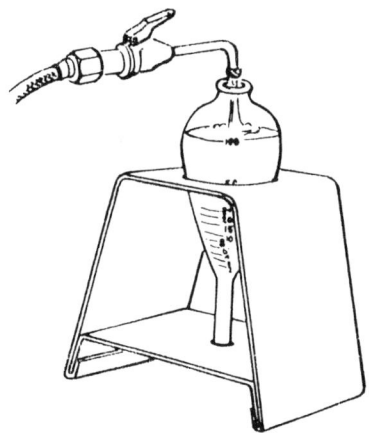

- Allow the centrifuge tube to stand undisturbed for a dwell time of at least 60 minutes for oil base mediums and 30 minutes for water base mediums.

Turn ahead to page 4-11.

From page 4-4

Absolutely. Magnetic particles would have a very low coercive force. Since the particles retain practically no residual magnetism, they are easily removed from the article if they are not held by a leakage field.

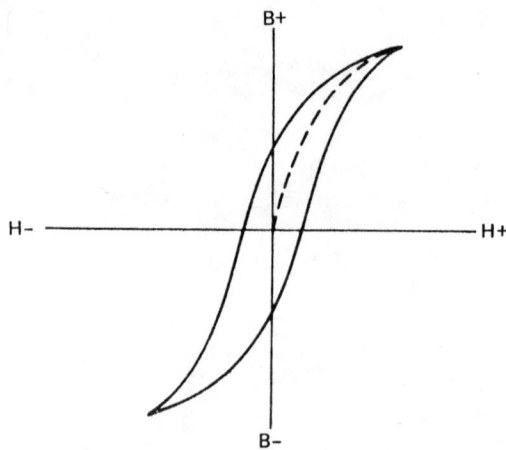

In summary, magnetic particles used in magnetic particle testing must:

- be highly permeable.
- have very low retentivity.
- have a very low coercive force.

The magnetic particles used in the wet bath medium are selected because of their size, shape, color, and magnetic properties. They are normally available as a dry powder. The powder is then easily mixed with the water or oil bath. When water baths are used, particles are mixed in a bath that will also contain rust inhibitors, wetting agents, and anti-foamers. The choice of particles to mix in the bath (water or oil) depends primarily upon which gives the better color contrast against the surface of articles to be tested. In general, black particles give a better contrast on new or finished articles, while red gives a better contrast on dark or used articles. Maximum contrast is usually obtained with fluorescent mediums.

Turn back to page 4-5.

From page 4-9                                                              4-11

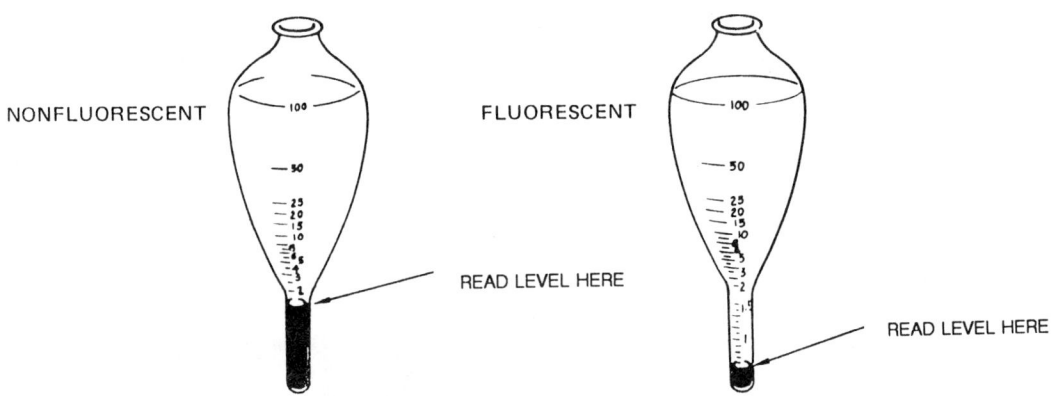

- Finally, read the volume of particles settled in the centrifuge tube. For nonfluorescent baths, the reading should be between 1.2 and 2.4 cc (ml). For fluorescent baths, the reading should be between 0.1 and 0.4 cc (ml).

If the reading is higher than indicated above, add water or oil depending on the medium used. If the reading is lower than indicated above, add particles to the bath.

**Note: Specifications or practices differ as to the acceptable volume of settled particles. The operator should always follow procedure.**

Turn to the next page.

Particles used in dry magnetic particle testing have similar characteristics to those used in baths except they remain in a dry, powder form. Color choice, as before, is determined by the greatest color contrast on the object to be magnetized.

Dry magnetic particles depend upon air to carry them to the surface of the article. Here a dry powder spray gun is being used.

The means of dispensing dry magnetic particles from air spray guns is to allow a light cloud of particles to float over a horizontal surface or adjacent to vertical, slanted, or overhead surfaces so that the particles may drift down or past the test specimen and be attracted to very weak leakage fields.

Turn to the next page.

Whether wet or dry magnetic particles are used, it is *absolutely essential that the articles* to be magnetically tested *are clean* and free of dirt, grease, oil, rust, and loose scale. If the articles are not clean, mobility of magnetic particles may be hindered to the extent that the particles may not be attracted to leakage fields. The removal of paint will depend on the test conditions and the requirements of the procedure being used.

If the article is not clean, a wet bath may run off an oily or greasy surface. Dirt, grease, oil, rust, and loose scale can also contaminate a wet bath. Dry particles will stick to a dirty surface. In addition, articles tested by the dry particle medium must also be dry, as the particles will stick to a damp or wet surface.

The processes involved in cleaning of the many new types of materials is a very large subject in itself. Many different processes are required. It is not our intent to explore this broad subject here. Rather, the intent is to *emphasize the great importance of proper cleaning of articles prior to magnetic particle testing.*

Turn to the next page.

## Sensitivity of Mediums

We have already established the fact that alternating current (AC) is the most effective current to use in finding *surface* discontinuities. This is true because AC tends to flow near the surface; therefore, AC creates the strongest magnetic field at the surface. Since it is acknowledged that AC is superior in finding surface discontinuities, we will confine our discussion here to the detection of *subsurface* discontinuities.

The following illustration compares the abilities of the various currents using both wet and dry magnetic particle testing mediums in finding *subsurface* discontinuities.

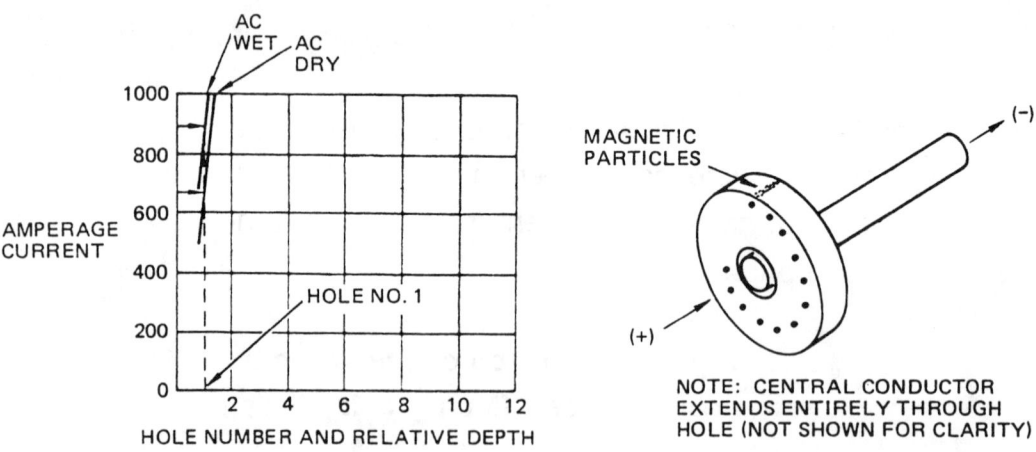

The above chart is based on tests made on a slice of cold-worked, tool-steel bar with holes drilled through it at varying depths below the surface.

Each test was made using a central conductor and the minimum amount of current of each type to produce a noticeable collection of magnetic particles on the outside surface of the article over any given hole.

Turn to the next page.

From page 4-14                                                                 4-15

With AC, using both wet and dry magnetic particles, about 700 amperes to 900 amperes were required to cause enough flux leakage to attract magnetic particles on the surface of the article in the vicinity of the No. 1 hole which is closest to the surface.

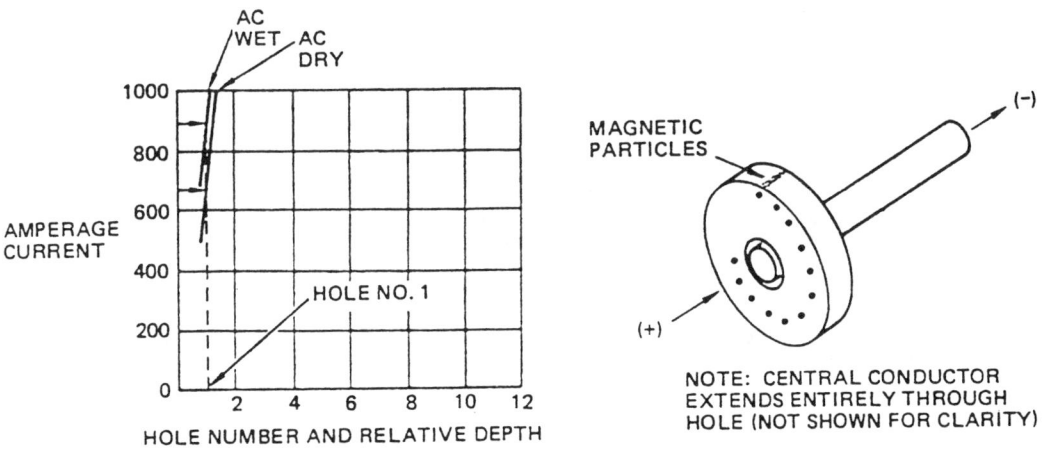

The closeness of the AC wet lines and AC dry lines in the graph indicates there is very little difference between the sensitivity of the two mediums. It also shows that alternating current is practically of no use in finding subsurface defects.

**In spite of its lack of penetration, you can tell from the two lines that alternating current would be most effective using which type medium?**

**Wet** .................................................. Page 4-16
**Dry** .................................................. Page 4-18

From page 4-15

You selected "wet." We will admit that those lines were very close together, but the AC dry line requires slightly less magnetizing current to obtain attraction at hole No. 1.

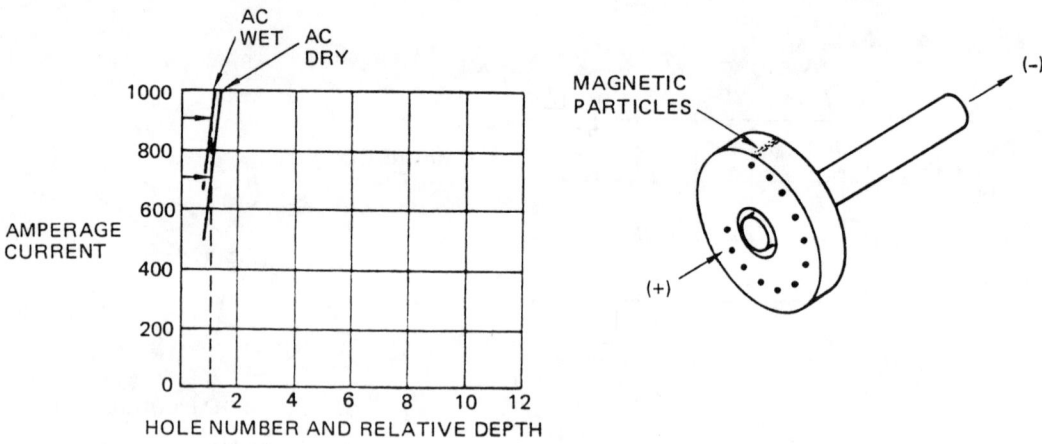

The arrow at about the 700 ampere point on the left scale shows the current required for the AC dry magnetic particles to be attracted to the leakage field created on the outside surface of the article over the first and most shallow hole. In other words, the AC dry line is to the right of the AC wet line which indicates that the dry particles were more easily attracted to a weaker leakage field. This results in better detection of subsurface discontinuities with the dry medium. The cyclic pulsing of the alternating current, plus the high mobility of the dry particles applied in a light cloud, allows the particles to be attracted with a lesser amount of AC.

Turn ahead to page 4-18.

From page 4-18                                                          4-17

You think that wet bath particles are more easily attracted to flux leakage. Let's look at that chart again and see if you won't change your mind.

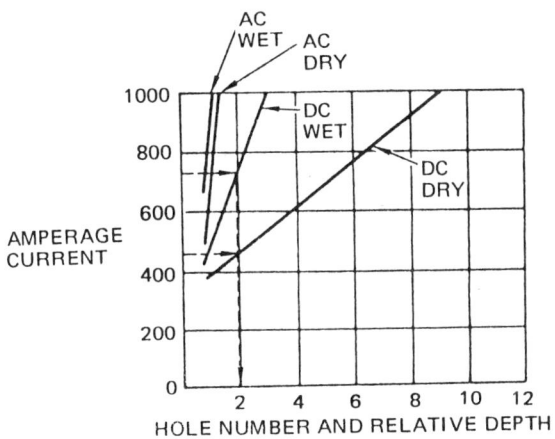

In the cases of both AC and (straight) DC, more amperage was required to create flux leakage when using the wet bath medium than when using dry magnetic particles. In other words, when using the wet bath medium, a stronger magnetic field was required to attract the wet bath particles. On the other hand, less current was required when the dry magnetic particles were used to obtain the same attraction. So you see, whether using AC or DC, dry magnetic particles are more easily attracted to flux leakage than are wet magnetic particles.

Turn ahead to page 4-20.

From page 4-15                                                                                   4-18

Right. Because the AC dry line is to the right of the AC wet line, you can see that a lesser amount of AC was required to cause enough flux leakage to attract dry magnetic particles on the surface of the article in the vicinity of the first (most shallow) hole. Now let us compare the use of wet and dry mediums when magnetizing with DC.

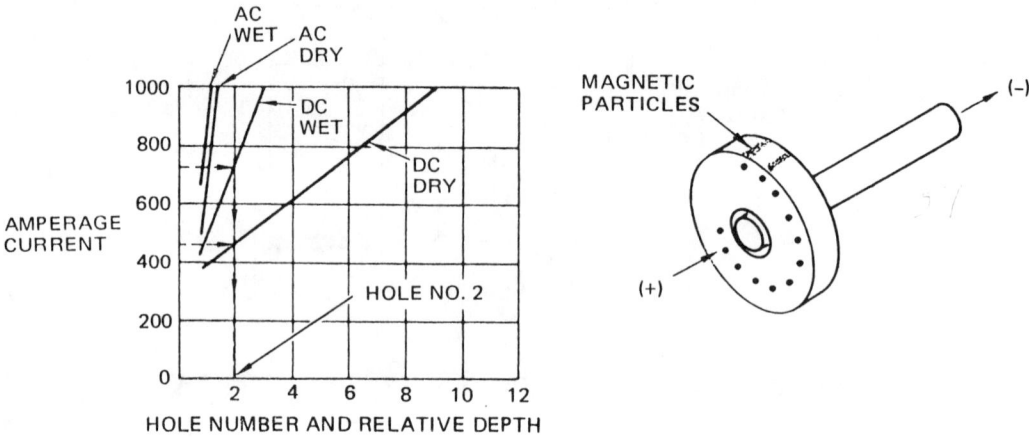

In comparing the two mediums, it is important to remember that the minimum amount of each type of current was used to obtain attraction of the magnetic particles. The wet bath medium using DC was able to attract magnetic particles on the surface over the second hole with a minimum current of about 735 amperes. With the use of dry magnetic particles and DC, only 475 amperes were required to attract magnetic particles on the surface at hole No. 2.

**From the above, we can conclude that whether AC or DC is used, which of the following is true?**

**Wet bath particles are more easily attracted to flux leakage** .................................................. **Page 4-17**
**Dry magnetic particles are more easily attracted to flux leakage** .................................................. **Page 4-20**

From page 4-20                                                          4-19

You think the DC wet medium is the most sensitive. Well, DC wet is more sensitive than either AC wet or AC dry, but there is a more sensitive medium. Let's look at the chart again.

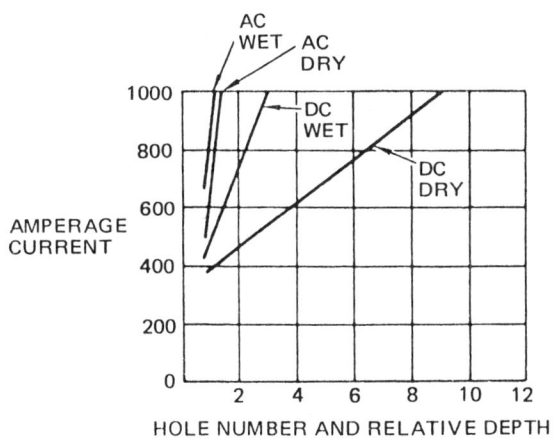

That should give you a clue to the correct answer.

Turn to the next page and try again.

From page 4-18

4-20

Absolutely. Whether using AC or DC, dry magnetic particles are more easily attracted to flux leakage. This is true because dry particles are blown in a cloud and allowed to drift gently to the part being magnetized. This allows the dry particles to be more easily attracted to weaker leakage fields; therefore, dry particles provide the greatest sensitivity.

With the wet bath medium using DC, 1000 amperes were required to cause enough flux leakage to attract magnetic particles at hole No. 3. With dry particles and DC, only about 550 amperes were required to attract particles at hole No. 3.

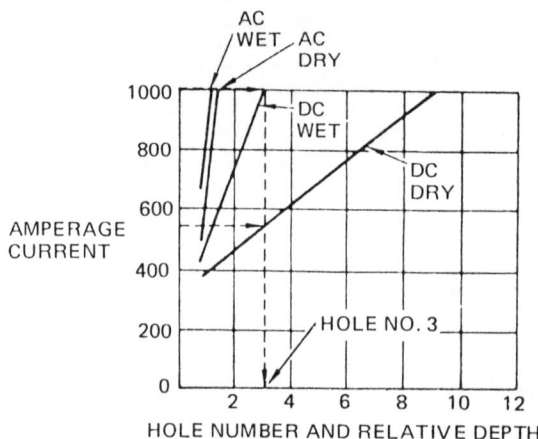
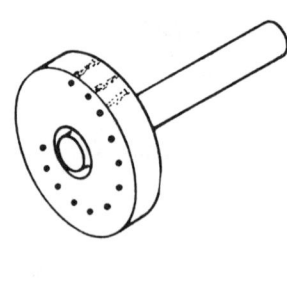

**Since dry powder magnetic particles are more easily attracted to weak leakage fields, we can say that they are more sensitive. Which medium is most sensitive up to this point?**

**DC wet** ................................................... Page 4-19
**DC dry** ................................................... Page 4-22
**AC dry** ................................................... Page 4-24

From page 4-22                                                              4-21

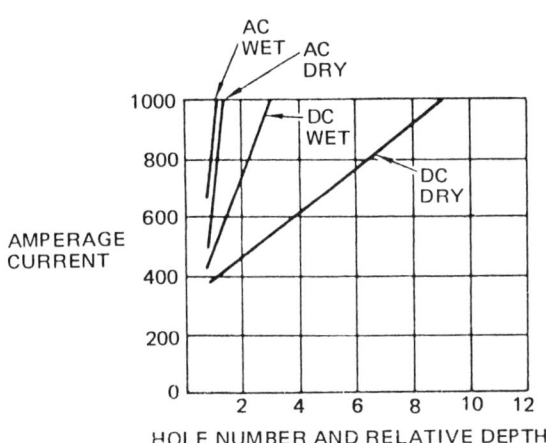

To show the greater sensitivity of DC using dry powder particles, how much amperage would be required for the DC dry medium to attract the particles to hole No. 6?

Approximately 1000 amperes  ..................... Page 4-23
Approximately 775 amperes   ..................... Page 4-25
Approximately 600 amperes   ..................... Page 4-27

Right you are. Straight DC with dry magnetic particles is the most sensitive medium up to this point. You seem to have the idea. One thing to remember is that the drilled holes in the test article are further and further away from the surface and that an increasingly higher amperage is required to cause a leakage field on the outside surface of the article.

The points to remember are:

- *Dry magnetic particles are more sensitive* than particles used in the wet bath medium whether AC or DC is used.

- *AC is most effective for locating surface defects.* AC is not effective in locating subsurface defects.

- *DC using dry powder particles is much more sensitive* than DC with the wet bath medium.

Turn back to page 4-21.

From page 4-21

You think it would take approximately 1000 amperes for the DC dry medium to attract magnetic particles to hole No. 6. OK, let's plot that on the chart and see if you are right.

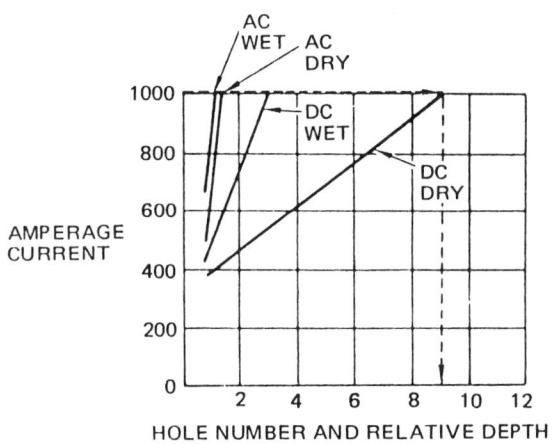

With 1000 amperes, DC using dry powder particles would attract magnetic particles to all holes up to and including No. 9. Remember, we want to use only the *minimum amount of current required* to cause flux leakage at an individual hole on the surface. In this case, we want to attract magnetic particles over hole No. 6.

Turn back to page 4-21 and try again.

From page 4-20                                                              4-24

You think AC dry is the most sensitive medium. Admittedly dry powder particles are the most sensitive when used with either AC or DC. But remember that we are now talking about *subsurface* discontinuities, and AC is used only for locating surface discontinuities.

Remembering that *dry powder particles are always more sensitive* than wet bath particles, turn back to page 4-20 and select the correct answer.

From page 4-21                                                                  4-25

Yes, of course. Approximately 775 amperes would be required to attract the dry particles to the outside surface in the vicinity of hole No. 6 using the DC dry medium.

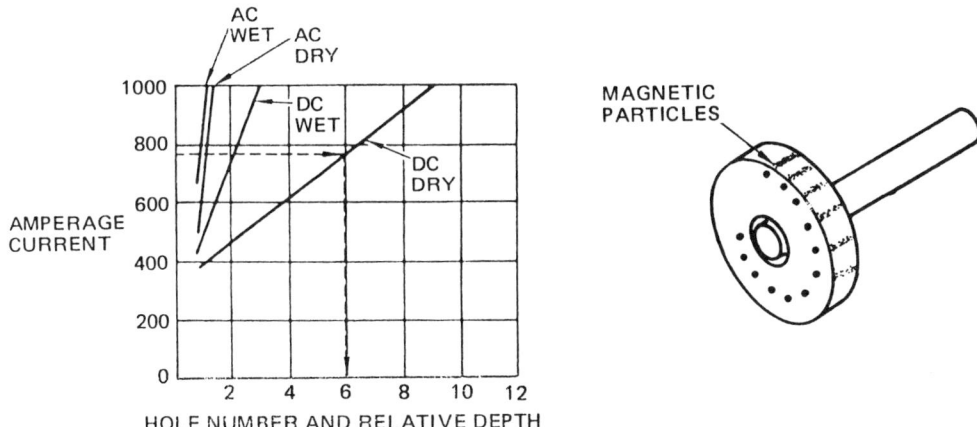

Notice also that the further the hole is from the surface, the "fuzzier" and less clearly defined the accumulation of magnetic particles becomes at the surface of the article.

Turn to the next page.

Now let's see where half-wave DC (HWDC) and full-wave DC (FWDC) fit into the picture.

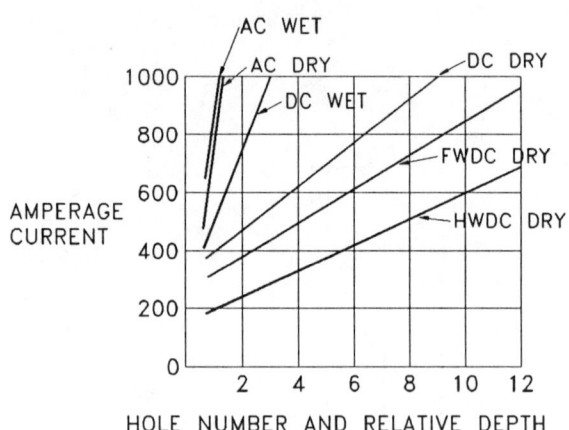

Here you can see that when using dry magnetic particles, the HWDC line would require only slightly more than 400 amperes to create flux leakage at hole No. 6 in the test article. The conclusion to be drawn here is that *HWDC has the greatest penetrating qualities*. Because of its continuous pulsing action, HWDC also agitates the magnetic particles which tends to give mobility to the particles. In this way, the magnetic particles can be attracted to very weak leakage fields.

Though we've not stressed it, FWDC responds similarly to the straight DC we've been discussing. However, the sensitivity of dry particle FWDC falls about halfway between the straight DC curve and the HWDC curve (for dry particles) illustrated above.

Turn ahead to page 4-28 for a brief review of this chapter.

From page 4-21

You think it would take approximately 600 amperes for the DC dry medium to attract magnetic particles to hole No. 6. Fine, let's plot it on the chart and see if you are right.

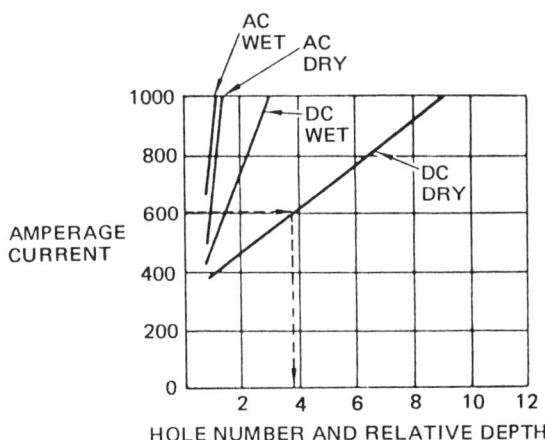

With this plot, you can see that 600 amperes using the DC dry medium wouldn't even cause flux leakage at hole No. 4.

Turn back to page 4-21 and boost the amperage a bit.

## CHAPTER REVIEW

__A__  1.  Magnetic particles are classed in accordance with the medium used, either wet or:

   A.   dry.
   B.   AC bath.
   C.   DC.
   D.   HWDC bath.

__C__  2.  The desired qualities of magnetic particles are shown by this hysteresis loop. The loop shows that the particles have a very _____ coercive force.

   A.   retentive
   B.   high
   C.   low
   D.   subsurface

__D__  3.  The magnetic particles retain very little residual magnetism which means they have very low:

   A.   permeability.
   B.   coercivity.
   C.   flux leakage.
   D.   retentivity.

From page 4-28

__C__  4. Wet magnetic particles are suspended in a liquid bath while dry particles are carried to the surface of an article by:

    A. humidity.
    B. a liquid flux.
    C. air.
    D. a tube.

__A__  5. Whether alternating current (AC) or direct current (DC) is used, dry magnetic particles are more easily attracted to flux:

    A. leakage.
    B. procedures.
    C. fluorescence.
    D. holes.

__C__  6. Since dry magnetic particles are more easily attracted to weak leakage fields, we can say they are more:

    A. fluorescent.
    B. brilliant.
    C. sensitive.
    D. highly refined.

___B___  7.  Straight DC using dry particles is more sensitive than when using DC with wet:

   A.   yokes.
   B.   particles.
   C.   towels.
   D.   residual fields.

___A___  8.  Magnetic particles used in either the wet or dry medium must be very easy to magnetize so they are highly:

   A.   permeable.
   B.   coercive.
   C.   fluorescent.
   D.   concentrated.

___D___  9.  Dry magnetic particles are blown in a light cloud so they drift slowly to the article being magnetized. For this reason, the particles are more mobile and easily attracted to weak leakage:

   A.   powders.
   B.   mediums.
   C.   fronts.
   D.   fields.

__A__ 10. Which technique and medium is most effective for location of surface discontinuities?

   A. AC with dry powder
   B. AC with the wet bath
   C. DC with wet bath

__C__ 11. Which technique and medium is most effective for location of deep subsurface discontinuities?

   A. Straight DC with dry powder
   B. AC with dry powder particles
   C. HWDC with dry powder particles

__B__ 12. The _____ of the test article's surface is of utmost importance in magnetic particle testing.

   A. smoothness
   B. cleanliness
   C. color
   D. shape

Turn to the next page for answers to these review questions.

## ANSWERS TO REVIEW QUESTIONS
## FOR CHAPTER 4

| Question & Answer | | Reference Page(s) |
|---|---|---|
| 1. | A | 4-1 |
| 2. | C | 4-4, 4-10 |
| 3. | D | 4-4 |
| 4. | C | 4-12 |
| 5. | A | 4-18, 4-20 |
| 6. | C | 4-22 |
| 7. | B | 4-22 |
| 8. | A | 4-10 |
| 9. | D | 4-12 |
| 10. | A | 4-22 |
| 11. | C | 4-26 |
| 12. | B | 4-13 |

Turn to the next page.

From page 4-32

Congratulations! You have just completed Volume I of the programmed instruction course of Magnetic Particle Testing.

Now you may want to evaluate your knowledge of the material presented in this handbook. A set of self-test questions is included at the back of this book. The answers can be found at the end of the test.

We want to emphasize that the test is for *your own* evaluation of *your* knowledge of the subject. If you elect to take the test, be honest with yourself—don't refer to the answers until you have finished. Then you will have a meaningful measure of your knowledge.

A score of 80% or better is considered typical, although your score may vary depending on your background and experience. If you find that you have trouble in some part of the test, it is up to you to review the material until you are satisfied that you know it.

Find a convenient time during which you can dedicate an hour or so and take the self-test. Every attempt should be made to complete the test in one sitting.

Turn to page A-1 and begin the test.

# APPENDIX A

# MAGNETIC PARTICLE TESTING

## SELF-TEST

__B__  1.  Which of the following correctly states the magnetic law of attraction?

    A. Like poles attract each other; opposite poles repel each other.
    B. Like poles repel each other; opposite poles attract each other.

__D__  2.  In what direction is the electric current flowing through this round steel bar?

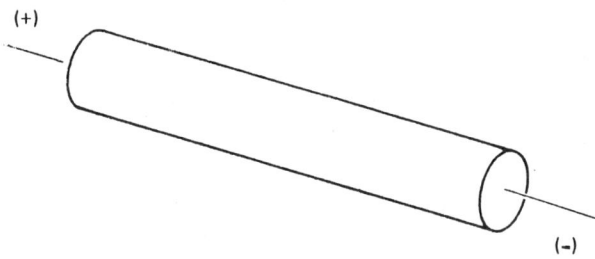

    A. Counterclockwise around the bar
    B. Right to left
    C. Clockwise around the bar
    D. Left to right

C   3.  What is the direction of the magnetic field in this round steel bar? (Assume that electric current is flowing through the bar.)

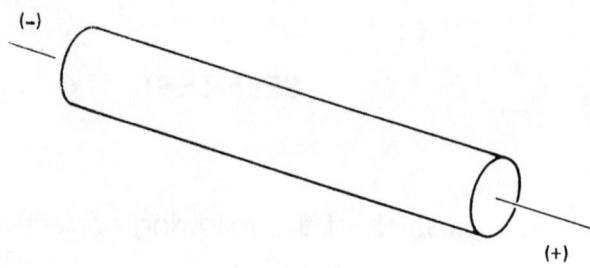

    A.  Left to right
    B.  Right to left
    C.  Clockwise around the bar
    D.  Counterclockwise around the bar

D   4.  Magnetic particle inspection is a means for detecting discontinuities at or near the surface in _____ materials.

    A.  metallic
    B.  nonconductive
    C.  ceramic
    D.  ferromagnetic

C   5.  The ease with which materials can be magnetized is called:

    A.  coercive force.
    B.  reluctance.
    C.  permeability.
    D.  flux density.

A-2

_A_  6.  The ability of material to retain a certain amount of residual magnetism is the definition of:

   A. retentivity.
   B. coercive force.
   C. flux density.
   D. permeability.

_B_  7.  The magnetic field which remains in a material after the magnetizing force is removed is called:

   A. permeability.
   B. residual magnetism.
   C. flux density.
   D. coercive force.

_D_  8.  The reverse magnetizing force required to remove residual magnetism from the material is referred to as:

   A. flux density.
   B. residual magnetism.
   C. retentivity.
   D. coercive force.

_D_  9.  The resistance of a material to changes in magnetic field strength is called:

   A. permeability.
   B. coercive force.
   C. resistance.
   D. reluctance.

A-3

___C___  10.  The number of lines of force per unit area is termed:

   A. retentivity.
   B. coercive force.
   C. flux density.
   D. permeability.

___B___  11.  If a ferromagnetic article is placed in a coil and the current turned on, the article will:

   A. get hot.
   B. be attracted to the coil.
   C. fluoresce.
   D. become diamagnetic.

___B___  12.  Ferromagnetic material placed within 18 inches (45 cm) of an energized coil will become adequately magnetized for magnetic particle inspection.

   A. True
   B. False

___D___  13.  Magnetization in which an external field is likely to be present in the article is _____ magnetization.

   A. circular
   B. unwanted
   C. prod
   D. yoke

A-4

Below are two hysteresis loops—one for soft iron and one for hardened ferromagnetic steel. Analyze each loop and fill in the blank spaces of the five items below each loop with "A" for high or "B" for low.

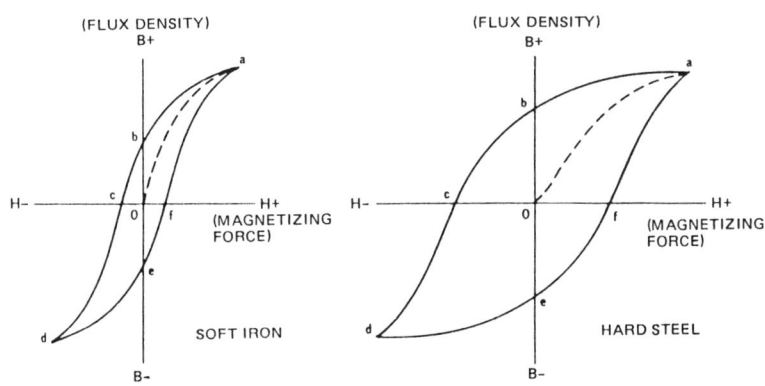

This loop shows:                          This loop shows:

14. __A__ permeability              19. __B__ permeability
15. __B__ retentivity               20. __A__ retentivity
16. __B__ coercive force            21. __A__ coercive force
17. __B__ reluctance                22. __A__ reluctance
18. __B__ residual magnetism        23. __A__ residual magnetism

__C__  24. Longitudinal magnetization can be established in an article by:

   A. passing current through a central conductor.
   B. passing current through the length of the article.
   C. placing the part inside a coil through which current is flowing.
   D. placing the part outside of a coil through which current is flowing.

A-5

B  25. The two primary techniques of inducing circular magnetism in an article are:

   A. with a coil and a central conductor.
   B. with a central conductor and by placing the article between the heads.
   C. with a central conductor and by use of a yoke.
   D. with a central conductor and by placing the article on top of the heads.

C  26. When a hollow tubular part is magnetized between the heads, flux density will be greatest:

   A. at the inside surface (ID).
   B. at the center of the hole in the article.
   C. at the outside surface (OD).
   D. Cannot be determined.

A  27. When using a central conductor to magnetize a hollow article, flux density in the article will be greatest:

   A. at the inside surface of the article.
   B. at the outside surface of the article.
   C. Both of the above.
   D. Neither of the above.

C  28. The maximum spacing between prods for the best sensitivity is:

   A. 2 inches (5 cm).
   B. 4 inches (10 cm).
   C. 8 inches (20 cm).
   D. 12 inches (30 cm).

29. The amount of current to be applied when using prods is based on the distance between the prods and:

   A. the thickness of the article.
   B. the L/D ratio of the article.
   C. the expected discontinuities in the article.
   D. the D/L ratio of the article.

30. A circular magnetic field is the nearest approach to a "leakage free" magnetic field.

   A. True
   B. False

31. What is an advantage of using a central conductor to magnetize articles?

   A. Detection of cracks on the outside surface of hollow articles
   B. Detection of laminations
   C. Detection of cracks on the inside
   D. None of the above.

32. A material that possesses the ability to attract iron and other ferromagnetic materials is generally called a:

   A. central conductor.
   B. magnet.
   C. discontinuity.
   D. flux capacitor.

A-7

A    33. The space around a magnet within which ferromagnetic materials are attracted to the magnet is called:

   A. a magnetic field.
   B. permeability.
   C. a hysteresis curve.
   D. a magnetic pole.

D    34. The magnetic field surrounding a bar magnet is most dense:

   A. near the middle of the magnet.
   B. 10 inches (25 cm) from the magnet.
   C. 3 inches (8 cm) from the magnet.
   D. at the ends of the magnet.

D    35. A material which retains its magnetism once it has been magnetized is called a _____ magnet.

   A. manufactured
   B. temporary
   C. bilateral
   D. permanent

B    36. When a round steel bar is magnetized by passing alternating current through its length, flux density is:

   A. greatest along its center line.
   B. greatest at the surface.
   C. uniform throughout its cross section.
   D. greatest at the ends of the material.

A-8

37. In magnetic particle testing, circular magnetization is obtained by:

   A. placing test articles within a current-carrying coil.
   B. using a yoke magnet with poles displaced along the length of articles.
   C. passing current directly through the length of the article.
   D. using a residual magnet.

38. The length of an article being magnetized between the heads has little effect on the strength of the magnetic field produced.

   A. True
   B. False

39. If a current of the same amperage is passed through two conductors, one of which is magnetic and the other nonmagnetic, the magnetic conductor will have a stronger field within it than will the nonmagnetic conductor.

   A. True
   B. False

__D__  40. The strength of the magnetic field within a coil is determined by:

   A. the current in the coil.
   B. the number of turns in the coil.
   C. the diameter of the coil.
   D. All of the above.

__C__  41. In magnetic particle testing, *best* results are obtained when the magnetic field is:

   A. straight DC.
   B. so that the magnetic field parallels the direction of the discontinuity.
   C. in a direction crosswise to the direction in which the discontinuity lies.

__B__  42. Surface seams in rolled bars are best detected in magnetic particle testing with _____ magnetization.

   A. yoke
   B. circular
   C. prod
   D. longitudinal

__A__  43. Reliable testing in the wet method depends on the equipment and the bath being:

   A. free of contaminants.
   B. adaptable to both water and oil.
   C. fluorescent.
   D. hot.

A-10

B   44.  A 60 hertz alternating current tends to flow through an article:

   A. evenly throughout the cross section of the article.
   B. near the surface of the article.
   C. near the inside surface of a hollow article.
   D. None of the above.

B   45.  One of the reasons half-wave DC (HWDC) is more sensitive than true direct current is because of its ability to:

   A. demagnetize the magnetic particles.
   B. agitate the magnetic particles.
   C. cause a stronger field at the surface of the article.
   D. All of the above.

C   46.  Which type of current is best for detection of surface discontinuities?

   A. Direct current
   B. Half-wave DC
   C. Alternating current
   D. Full-wave DC

B   47.  Alternating current can be used to locate subsurface discontinuities.

   A. True
   B. False

A-11

__A__ 48. Whether using AC or DC, which type of magnetic particles provides the greatest sensitivity?

A. Dry magnetic particles
B. Particles used in the wet bath method
C. Red particles
D. Fluorescent particles

__D__ 49. Which method is most effective for locating deep subsurface discontinuities?

A. Straight DC with dry powder
B. AC with dry powder
C. AC with wet particles
D. HWDC with dry powder

__C__ 50. A test object 10 inches (25 cm) long and 2 inches (5 cm) in diameter, longitudinally magnetized in a 5-turn coil, requires a magnetizing current of _____ amps.

A. 600
B. 1200
C. 1800
D. 2400

__B__ 51. The strength (concentration of particles) in a bath must remain constant if the _____ of identical discontinuities is not to vary.

A. reluctance
B. indications
C. magnetic field
D. depth

52. Which of the following methods is the more sensitive?

    A. Bilateral
    B. Residual
    C. Luxation
    D. Continuous

53. When using a coil, what is the effective distance of the magnetic field?

    A. 6 to 9 inches (15 to 23 cm) on each side of the coil
    B. 18 inches (46 cm) on each side of the coil
    C. 6 to 9 inches (15 to 23 cm)
    D. 6 to 18 inches (15 to 46 cm)

54. Which cracks in this round bar can be detected by longitudinal magnetization?

    A. A
    B. B and C
    C. A and B
    D. A and C

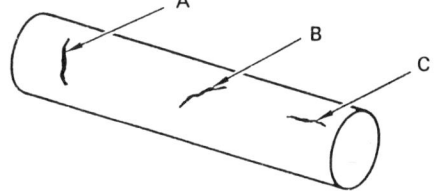

55. Which cracks in this round bar can be detected by circular magnetization?

    A. A
    B. A and B
    C. B and C
    D. A and C

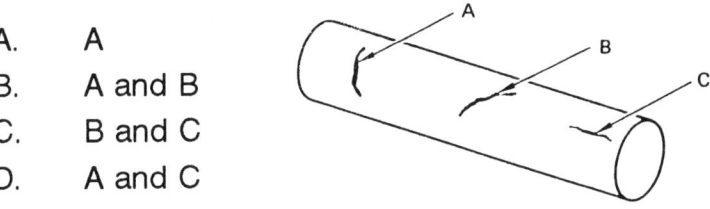

Insert in the blank in front of each letter the best magnetizing technique of locating each discontinuity.

A. circular between the heads
B. circular with central conductor
C. longitudinal in coil
D. A and B
E. A and C
F. B and C

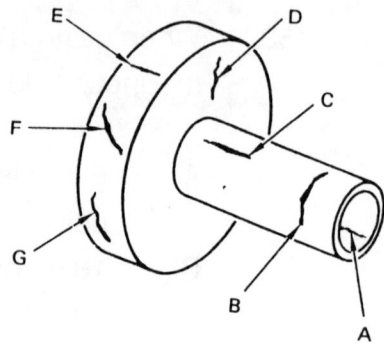

56. __B__ Discontinuity A
57. __C__ Discontinuity B
58. __D__ Discontinuity C
59. __D__ Discontinuity D
60. __B__ Discontinuity E
61. __E__ Discontinuity F
62. __C__ Discontinuity G

__C__ 63. _____ amperes of current would be required to circularly magnetize an article 10 inches (25 cm) long and 2 inches (5 cm) in diameter.

700 to 1,000 ampers per Inch of Dia.

A. 2800 to 4000
B. 2100 to 3000
C. 1400 to 2000
D. 700 to 1000

A-14

D  64. To obtain best results when using prods, what should be the distance between prods?

   A. 5 to 10 inches (13 to 25 cm)
   B. 4 to 6 inches (10 to 15 cm)
   C. 12 to 16 inches (30 to 41 cm)
   D. 6 to 8 inches (15 to 20 cm)

B  65. _____ coil shots would be required to adequately magnetize a bar 24 inches (61 cm) long.

   A. One
   B. Two
   C. Three
   D. Four

B  66. Flux density is greatest at what point on a coil?

   A. At the center of the coil
   B. At the inside surface of the coil
   C. Outside the coil
   D. In the center of the coil windings

C  67. If an article is being magnetized between the heads, where is the flux density greatest?

   A. In the center of the article
   B. Outside the article
   C. At the surface of the article

A-15

_C_ 68. When magnetizing with a central conductor, flux density is greatest:

    A. near the outside surface of the article being magnetized.
    B. outside the conductor.
    C. at the surface of the conductor.
    D. at the center of the conductor.

_C_ 69. In what way should dry magnetic particles be applied to a magnetized area?

    A. Poured on the area
    B. Blown forcibly at the area
    C. Allowed to drift to the area in a light cloud
    D. They shouldn't.

_A_ 70. The residual method of testing could be expected to give best test results on a _____ permeability material.

    A. low
    B. high
    C. medium
    D. zero

C̲ 71. When magnetizing with the central conductor, _____ amperes would be required to circularly magnetize a hollow tube with an outside diameter of 2 inches (5 cm).

   A. 350 to 600
   B. 700 to 1000
   C. 1400 to 2000
   D. 1750 to 3000

D̲ 72. Magnetic particle indications are caused when magnetic particles are attracted at which of the following?

   A. Flux leakage
   B. Magnetic poles
   C. Leakage fields
   D. All of the above

A-17

# ANSWERS FOR SELF-TEST

| Question & Answer | Reference Page(s) | Question & Answer | Reference Page(s) |
|---|---|---|---|
| 1. B | 1-3 | 21. A | 1-77 |
| 2. D | 2-1 | 22. A | 1-77 |
| 3. C | 2-2 | 23. A | 1-77 |
| 4. D | 1-1 | 24. C | 2-59 |
| 5. C | 1-45 | 25. B | 2-54, 2-55 |
| 6. A | 1-61 | 26. C | 2-31, 2-43 |
| 7. B | 1-53 | 27. A | 2-41 |
| 8. D | 1-63 | 28. C | 2-77, 3-35 |
| 9. D | 1-73 | 29. A | 3-35 |
| 10. C | 1-49 | 30. A | 2-11 |
| 11. B | 2-64 | 31. C | 2-41 |
| 12. B | 2-67, 2-70 | 32. B | 1-1 |
| 13. D | 2-81 | 33. A | 1-12 |
| 14. A | 1-78 | 34. D | 1-12 |
| 15. B | 1-78 | 35. D | 1-47 |
| 16. B | 1-78 | 36. B | 3-10 |
| 17. B | 1-78 | 37. C | 2-6 |
| 18. B | 1-78 | 38. A | 2-31 |
| 19. B | 1-77 | 39. A | 2-13 |
| 20. A | 1-77 | 40. D | 2-56 |

| Question & Answer | Reference Page(s) | Question & Answer | Reference Page(s) |
|---|---|---|---|
| 41. C | 2-18, 2-19 | 66. B | 2-56 |
| 42. D | 2-27 | 67. C | 2-31, 2-32 |
| 43. A | 4-5 | 68. C | 2-41 |
| 44. B | 3-2 | 69. C | 4-12 |
| 45. B | 3-9 | 70. A | 1-18, 1-19 |
| 46. C | 3-2 | 71. C | 3-33 |
| 47. B | 3-10 | 72. D | 1-25, 1-30 |
| 48. A | 4-20 | | |
| 49. D | 4-26 | | |
| 50. C | 3-37 | | |
| 51. B | 4-5 | | |
| 52. D | 4-27 | | |
| 53. A | 3-40 | | |
| 54. C | 2-57 | | |
| 55. C | 2-18, 2-59 | | |
| 56. B | 2-44 | | |
| 57. C | 2-57, 2-59 | | |
| 58. D | 2-44 | | |
| 59. D | 2-53 | | |
| 60. D | 2-44 | | |
| 61. E | 2-18, 2-59 | | |
| 62. C | 2-18, 2-59 | | |
| 63. C | 3-23 | | |
| 64. D | 2-77 | | |
| 65. B | 3-40 | | |

A-19

# APPENDIX B

# GLOSSARY

**Air Gap** When a magnetic circuit contains a small gap which the magnetic flux must cross, the space is referred to as an air gap. Cracks produce small air gaps on the surface of an article.

**Alternating Current** Electric current periodically reversing in polarity or direction of flow.

**Ampere** The unit of electrical current. One ampere is the current that flows through a conductor having a resistance of one ohm at a potential of one volt.

**Ampere Turns** The product of the number of turns in a coil and the number of amperes flowing through it. A measure of the magnetizing or demagnetizing strength of the coil.

**Arc Strike** Localized burn damage to a test article caused by making or breaking an energized electrical circuit.

**Background** The appearance of the test article's surface. Part of the viewing area when interpreting surface indications.

**Bath** The suspension of iron oxide particles in a liquid medium or vehicle (light oil or water).

**Black Light** Radiant energy in the near ultraviolet range. This light has a wavelength of 3200 to 4000 angstroms (Å), peaking at 3650 Å, on the spectrum. This is between visible light and ultraviolet light.

**Black Light Filter**  A filter that transmits black light while suppressing the transmission of visible light and harmful ultraviolet radiation.

**Carbon Steel**  Steel which does not contain significant amounts of alloying elements other than carbon and manganese.

**Central Conductor**  An electrical conductor that is passed through the opening in a ring or tube or any hole in an article for the purpose of creating a circular field in the ring or tube or around the hole.

**Circular Field**  See "**Field, Circular Magnetic.**"

**Circular Magnetization**  A method of inducing a magnetic field in an article so that the magnetic lines of force take the form of concentric rings about the axis of the current.  This is accomplished by passing the current directly through the article or through a conductor which passes into or through a hole in the article.  The circular method is applicable for the detection of discontinuities with axes approximately parallel to the axis of the current through the article.

**Coercivity or Coercive Force**  The strength of the magnetizing field necessary to reduce its residual (remanent) magnetism to zero.

**Coil Shot**  A pulse of magnetizing current passed through a coil surrounding an article for the purpose of longitudinal magnetization.

**Contact Head**  The electrode fixed to the magnetic particle testing unit through which the magnetizing current is drawn.

**Contact Pads**  Replaceable metal pads, usually of copper braid, placed on contact heads to give good electrical contact thereby preventing damage to the article under test.

**Continuous Technique** An inspection technique in which ample amounts of magnetic particles are applied or are present on the piece during the time the magnetizing current is applied.

**Core** That part of the magnetic circuit which is within the electrical winding.

**Curie Point** The temperature at which ferromagnetic materials can no longer be magnetized by outside forces and at which they lose their residual magnetism: approximately 1200°F to 1600°F (650°C to 870°C) for many metals.

**Current Flow Technique** A technique of circular magnetization in which current is passed through an article via prods or contact heads. The current may be alternating, half-wave rectified, rectified alternating, or direct.

**Current Induction Technique** A technique of magnetization in which a circulating current is induced in a ring-shaped component by a fluctuating magnetic field.

**Defect** A discontinuity that interferes with the usefulness of an article or exceeds acceptability limits established by applicable procedures or specifications.

**Demagnetization** The reduction in the degree of residual magnetism to an acceptable level.

**Diamagnetic** Materials whose atomic structure won't permit any real magnetization. Materials such as bismuth and copper are diamagnetic.

**Diffused Indications** Indications that are not clearly defined, such as indications of subsurface defects.

**Direct Contact Magnetization**  A magnetic particle testing technique in which current is passed through the test article. This includes head shots and prod shots.

**Direct Current**  An electrical current which flows steadily in one direction.

**Discontinuity**  An interruption (cracks, forging laps, seams, inclusions, porosity, etc.) in the normal physical structure of the configuration of an article. A discontinuity may or may not affect the usefulness of the article.

**Distorted Field**  A field that does not follow a straight path or have a uniform distribution. This occurs in irregularly-shaped objects.

**Dry Medium**  Magnetic particle inspection in which the particles employed are in the dry powder form.

**Dry Powder**  Finely-divided ferromagnetic particles suitably selected and prepared for magnetic particle inspection.

**Electromagnet**  A magnet created by inserting a suitable metal core within or near a magnetizing field formed by passing electric current through a coil of insulated wire.

**Etching**  The process of exposing subsurface conditions of metal articles by removal of the outside surface through the use of chemical agents. Due to the action of the chemicals in eating away the surface, various surface or subsurface conditions are exposed or exaggerated and made visible to the eye.

**False Indication**  Indications formed which are not a result of flux leakage.

**Ferromagnetic**  A term applied to materials which can be magnetized and strongly attracted by a magnetic field.

**Field, Circular Magnetic** Generally the magnetic field in and surrounding any electrical conductor or article resulting from a current being passed through the conductor or article or from prods.

**Field, Longitudinal Magnetic** A magnetic field wherein the flux lines traverse the component in a direction essentially parallel with the axis of the magnetizing coil or to a line connecting the two poles at the magnetizing yoke.

**Field, Magnetic** The space within and surrounding a magnetized article, or a conductor carrying current in which the magnetic force is present.

**Field, Magnetic Leakage** The magnetic field that leaves or enters the surface of an article at a magnetic pole.

**Field, Multidirectional** A magnetic field that is the result of two magnetic forces impressed upon the same area of a magnetizable object at the same time—sometimes called a "**vector field**."

**Field, Residual Magnetic** The field that remains in magnetizable material after the magnetizing force has been removed.

**Field, Vector** See **Field, Multidirectional**.

**Flash Magnetization** Magnetization by a current flow of very brief duration.

**Flaw** A detected discontinuity that is not necessarily cause for rejection of the test article.

**Fluorescence** The emission of visible radiation by a substance as the result of and only during the absorption of black light radiation.

**Fluorescent Magnetic Particle Inspection**  The magnetic particle inspection process employing a finely-divided fluorescent ferromagnetic inspection medium that fluoresces when activated by black light.

**Flux Density**  The normal magnetic flux per unit area. It is designated by the letter "B" and is expressed in telsa (SI units) or gauss (cgs units).

**Flux Leakage**  Magnetic lines of force which leave and enter an article at poles on the surface.

**Flux Lines**  Imaginary magnetic lines used as a means of explaining the behavior of magnetic fields. Their conception is based on the pattern of lines produced when iron filings are sprinkled over a piece of paper laid over a permanent magnet. Also called **Lines of Force**.

**Flux Penetration, Magnetic**  The depth to which a magnetic flux is present in an article.

**Furring**  Buildup or bristling of magnetic particles due to excessive magnetization of the article.

**Gauss**  The unit of flux density. Numerically, one gauss is one line of flux per square centimeter of area and is designated by the letter "B."

**Head Shot**  A short pulse of magnetizing current passed through an article or a central conductor while clamped between the head contacts of a stationary magnetizing unit for the purpose of circularly magnetizing the article.

**Heads**  The clamping contacts on a stationary magnetizing unit.

**Horseshoe Magnet**  A bar magnet bent into the shape of a horseshoe so that the two poles are adjacent. Usually the term applies to a permanent magnet.

**Hysteresis**  The lagging of the magnetic effect when the magnetic force acting upon a ferromagnetic body is changed; the phenomenon exhibited by a magnetic system wherein its state is influenced by its previous magnetic history.

**Hysteresis Loop**  A curve showing the flux density, "B," plotted as a function of magnetizing force, "H." As the magnetizing force is increased to the saturation point in the positive, negative, and positive direction sequentially, the curve forms a characteristic S-shaped loop. Intercepts of the loop with the "B" and "H" axes and the points of maximum and minimum magnetizing force define important magnetic characteristics of the material.

**Indication**  A magnetic particle accumulation on the surface of an article being tested.

**Inductance**  The magnetism produced in a ferromagnetic body by some outside magnetizing force. The magnetism is not the result of passing current through the article.

**Inspection**  The process of examining and checking materials and articles for possible defects or for deviation from established standards.

**Interpretation**  The determining of the cause and significance of indications of discontinuities from the standpoint of whether they are detrimental defects or false or nonrelevant indications.

**Leakage Field**  The magnetic field forced out into the air by the distortion of the field within an article.

**Light Intensity**  The light energy reaching a unit of surface area per unit of time.

**Lines of Force**  See **Flux Lines**.

**Longitudinal Field**  See **Field, Longitudinal Magnetic**.

**Longitudinal Magnetization**  The process of inducing a magnetic field into the article such that the magnetic lines of force extending through the article are approximately parallel to the axis of the magnetizing coil or to a line connecting the two poles when yokes (electromagnets) are used.

**Magnet, Permanent**  A highly-retentive metal that has been strongly magnetized; i.e., the alloy Alnico.

**Magnetic Field**  See **Field, Magnetic**.

**Magnetic Field Indicator**  An instrument designed to detect and/or measure the flux density and polarity of magnetic fields.

**Magnetic Field Strength**  The measured intensity of a magnetic field at a point always external to the magnet or conductor; usually expressed in amperes per meter or oersted (Oe).

**Magnetic Material**  Those materials that are attracted by magnetism.

**Magnetic Particles**  Finely divided ferromagnetic material.

**Magnetic Particle Inspection**  A nondestructive inspection method for locating discontinuities in ferromagnetic materials.

**Magnetic Poles**  Concentration of flux leakage in areas of discontinuities, shape changes, permeability variations, etc.

**Magnetic Writing**  A form of nonrelevant indications caused when the surface of a magnetized part comes in contact with another piece of ferromagnetic material that is magnetized to a different value.

**Magnetizing Current**  The flow of either alternating, rectified alternating, or direct current used to induce magnetism into the article being inspected.

**Magnetizing Force**  The magnetizing field applied to a ferromagnetic material to induce magnetization.

**Medium**  The fluid in which fluorescent and nonfluorescent magnetic particles are suspended to facilitate their application in the wet method.

**Near Surface Discontinuity**  A discontinuity not open to, but located near, the surface of a test article.

**Nondestructive Inspection**  See **Nondestructive Testing (NDT)**.

**Nondestructive Examination**  See **Nondestructive Testing (NDT)**.

**Nondestructive Evaluation**  See **Nondestructive Testing (NDT)**.

**Nondestructive Testing (NDT)**  The application of methods to determine the usefulness of a test article which do not destroy its intended usefulness or fitness for purpose.  This involves the detection, location, interpretation and evaluation of discontinuities.

**Nonrelevant Indication**  A magnetic particle indication due to a leakage field which is not caused by an actual discontinuity in the magnetized material, but by some other condition which does not affect the usefulness of the article. False indications are nonrelevant.

**Oersted**  A unit of field strength which produces magnetic induction and is designated by the letter "H."

**Paramagnetic**  Materials which are slightly affected by a magnetic field. Examples are chromium, manganese, aluminum and platinum. A small group of these materials are classified as ferromagnetic.

**Permeability**  The ease with which the lines of force are able to pass through an article.

**Pole**  The area on a magnetized article from which the magnetic field is leaving or returning to the article.

**Prods**  Hand-held electrodes attached to cables used to transmit the magnetizing current from the source to the article under inspection.

**Rectified Alternating Current**  Alternating current which has been converted into direct current.

**Reluctance**  The resistance of a magnetic material to changes in magnetic field strength.

**Residual Field**  See **Field, Residual Magnetic**.

**Residual Magnetism**  The amount of magnetism that a magnetic material retains after the magnetizing force is removed. Also called "residual field" or "remanence."

**Residual Technique** A procedure in which the indicating material is applied after the magnetizing force has been discontinued.

**Resultant Field** See **Field, Resultant Magnetic**.

**Retentivity** The ability of a material to retain a certain portion of residual magnetization. Also known as remanence.

**Saturation** The point at which increasing the magnetizing force produces no further magnetism in a material.

**Sensitivity** The capacity or degree of responsiveness to magnetic particle inspection.

**Settling Test** A procedure used to determine the concentration of magnetic particles in a medium or vehicle.

**Skin Effect** The description given to alternating current magnetization due to its containment to the surface of a test article.

**Solenoid (Coil)** An electric conductor formed into a coil often wrapped around a central core of highly-permeable material.

**Subsurface Discontinuity** Any discontinuity which does not open onto the surface of the article in which it exists.

**Suspension** The correct term applied to the liquid bath in which the ferromagnetic particles used in the wet magnetic particle inspection method are suspended.

**Test Article** An article containing known artificial or natural defects used for checking the efficiency of magnetic particle flaw detection processes.

**Vector Field**  See **Field, Resultant Magnetic**.

**Wet Medium**  An inspection employing ferromagnetic particles suspended in a liquid (oil or water) as a vehicle.

**Yoke**  A U-shaped or C-shaped piece of highly-permeable magnetic material, either solid or laminated, sometimes with adjustable pole pieces (legs) around which is wound a coil carrying the magnetizing current.

**Yoke Magnetization**  A longitudinal magnetic field induced in an article or in an area of an article by means of an external electromagnet shaped like a yoke.

# APPENDIX C

# MEASUREMENT CONVERSION CHARTS

## U.S. CUSTOMARY TO INTERNATIONAL SYSTEM (SI) UNITS

| Inch to Millimeter Conversions | | | | | | | |
|---|---|---|---|---|---|---|---|
| Inch | Millimeter | Inch | Millimeter | Inch | Millimeter | Inch | Millimeter |
| 1 | 25.4 | 26 | 660.4 | 51 | 1295.4 | 76 | 1930.4 |
| 2 | 50.8 | 27 | 685.8 | 52 | 1320.8 | 77 | 1955.8 |
| 3 | 76.2 | 28 | 711.2 | 53 | 1346.2 | 78 | 1981.2 |
| 4 | 101.6 | 29 | 736.6 | 54 | 1371.6 | 79 | 2006.6 |
| 5 | 127.0 | 30 | 762.0 | 55 | 1397.0 | 80 | 2032.0 |
| 6 | 152.4 | 31 | 787.4 | 56 | 1422.4 | 81 | 2057.4 |
| 7 | 177.8 | 32 | 812.8 | 57 | 1447.8 | 82 | 2082.8 |
| 8 | 203.2 | 33 | 838.2 | 58 | 1473.2 | 83 | 2108.2 |
| 9 | 228.6 | 34 | 863.6 | 59 | 1498.6 | 84 | 2133.6 |
| 10 | 254.0 | 35 | 889.0 | 60 | 1524.0 | 85 | 2159.0 |
| 11 | 279.4 | 36 | 914.4 | 61 | 1549.4 | 86 | 2184.4 |
| 12 | 304.8 | 37 | 939.8 | 62 | 1574.8 | 87 | 2209.8 |
| 13 | 330.2 | 38 | 965.2 | 63 | 1600.2 | 88 | 2235.2 |
| 14 | 355.6 | 39 | 990.6 | 64 | 1625.6 | 89 | 2260.6 |
| 15 | 381.0 | 40 | 1016.0 | 65 | 1651.0 | 90 | 2286.0 |
| 16 | 406.4 | 41 | 1041.4 | 66 | 1676.4 | 91 | 2311.4 |
| 17 | 431.8 | 42 | 1066.8 | 67 | 1701.8 | 92 | 2336.8 |
| 18 | 457.2 | 43 | 1092.2 | 68 | 1727.2 | 93 | 2362.2 |
| 19 | 482.6 | 44 | 1117.6 | 69 | 1752.6 | 94 | 2387.6 |
| 20 | 508.0 | 45 | 1143.0 | 70 | 1778.0 | 95 | 2413.0 |
| 21 | 533.4 | 46 | 1168.4 | 71 | 1803.4 | 96 | 2438.4 |
| 22 | 558.8 | 47 | 1193.8 | 72 | 1828.8 | 97 | 2463.8 |
| 23 | 584.2 | 48 | 1219.2 | 73 | 1854.2 | 98 | 2489.2 |
| 24 | 609.6 | 49 | 1244.6 | 74 | 1879.6 | 99 | 2514.6 |
| 25 | 635.0 | 50 | 1270.0 | 75 | 1905.0 | 100 | 2540.0 |

To convert millimeters to centimeters, move decimal point one place to the left (e.g., 635 mm - 63.5 cm)

To convert millimeters to meters, move decimal point three (3) places to the left (e.g., 635 mm - 0.635m)

| Common Fractions to Millimeter Conversions ||
|---|---|
| Inch | Millimeters |
| 1/64 | 0.397 |
| 1/32 | 0.794 |
| 1/16 | 1.588 |
| 1/8 | 3.175 |
| 1/4 | 6.350 |
| 1/2 | 12.700 |
| 1 | 25.400 |